Number Categories

Studia Typologica

———

Beihefte / Supplements
STUF – Sprachtypologie und Universalienforschung
      Language Typology and Universals

Volume 32

Universität
Bremen

Number Categories

—

Dynamics, Contact, Typology

Edited by
Deborah Arbes

DE GRUYTER
MOUTON

ISBN 978-3-11-221483-1
e-ISBN (PDF) 978-3-11-098660-0
e-ISBN (EPUB) 978-3-11-098662-4
ISSN 1617-2957

Library of Congress Control Number: 2023936389

Bibliographic information published by the Deutsche Nationalbibliothek
The Deutsche Nationalbibliothek lists this publication in the Deutsche Nationalbibliografie;
detailed bibliographic data are available on the Internet at http://dnb.dnb.de.

© 2025 Walter de Gruyter GmbH, Berlin/Boston
This volume is text- and page-identical with the hardback published in 2023.
Cover image: Alpha-C/iStock/Thinkstock
Typesetting: Integra Software Services Pvt. Ltd.
Printing and binding: CPI books GmbH, Leck

www.degruyter.com

Preface

The authors of this volume came together on June 3rd 2021 in an online pre-conference workshop for the 4th Diversity Linguistics conference, organised by the University of Bremen. The topic of the workshop was "number categories" and we heard ten presentations about it altogether, five of which formed the basis for essays published in this volume.

Number is an essential grammatical category in languages worldwide. The ways in which number is marked on nominals can vary considerably from language to language. They can also change over time and be influenced by contact with other languages. These aspects were central in our talks and discussions and ultimately led to the title of this volume.

In this issue, we explore grammatical number from various angles: Katharina Gayler, Elsadig Omda Ibrahim Elnur and Isabel Compes investigate how number marking can be described in a tonal language: the Wagi dialect of Beria. Deborah Arbes and Alexandre Arkhipov analyse number marking in connection with language contact in two very different languages: Kamas (in contact with Russian) and Welsh (in contact with English). Typological perspectives are taken by Thomas Stolz and Silva Nurmio, who discover co-plurals and singulatives respectively.

Together, the articles of this volume offer a diverse portrait of the possibilities and complexities of number marking. We hope that this volume will not only contribute to our understanding of grammatical number and its categories, but also inspire further research in this area.

As the editor, I would like to thank all the authors who contributed to this volume, as well as the reviewers, who provided valuable feedback. Finally, I would like to thank Julia Nintemann, Nataliya Levkovych, Thomas Stolz and everyone else who has participated in and supported the 4th Diversity Linguistics conference, which provided a fertile ground for stimulating discussions and papers.

Deborah Arbes, Bremen, March 2023

https://doi.org/10.1515/9783110986600-202

Contents

Katharina Gayler, Elsadig Omda Ibrahim Elnur, and Isabel Compes

1 The tonal marking of number on nominals in the Wagi dialect of Beria

Abstract: This paper presents an analysis of the tonal marking of number on nominals in the Wagi dialect of Beria (Nilo-Saharan, Sudan). The investigation is based on archive data and fieldwork taking a descriptive approach. Seven nominal tone classes of distinct tonal patterns on singular and plural forms are identified. Singular tone patterns are analysed as lexically specified, plural tone patterns as combinations of the lexical pattern and a tonal plural morpheme. In noun phrases, this plural morpheme attaches only once at the right edge, which is in line with the behaviour of other grammatical markers in Wagi.

Keywords: noun phrase, nominals, tone classes, lexical tone, grammatical tone, plural marking, Beria

1 Introduction

This exploratory investigation gives a first account of number marking on isolated nominal forms and within noun phrases in the Wagi dialect of Beria (Nilo-Saharan, Sudan). Beria (also referred to by its exonym Zaghawa in the literature) is the only surviving Eastern-Saharan language and spoken in Northern Darfur at the border area of Chad and Sudan. There are four major dialects: Wagi, Kube, Tuba and Dirong-Guruf (Anonby & Johnson (2001: 9)). The present paper is concerned with the Wagi dialect as spoken in Sudan, where Wagi speakers are the biggest

Acknowledgements: The paper summarises revised parts of the MA thesis of Katharina Gayler (2021) and incorporates first results from the PhD project of Elsadig Omda Ibrahim Elnur on the tone system of Beria. Katharina Gayler is funded by the SFB 1252 'Prominence in Language', University of Cologne. Omda's PhD project is funded by the DAAD and situated at the Institute of Linguistics, University of Cologne. Isabel Compes was funded by the Institute of Linguistics, University of Cologne. Our thanks go to Elsadig Omda Ibrahim Elnur, Adam Rahma Jamous Mohamadeen and all Beria speakers in Sudan. We are very grateful to Gertrud Schneider-Blum and our reviewers Alexandre Arkhipov and Deborah Arbes for their critical comments on earlier versions of the paper which notably improved our presentation. Any remaining errors in the article are of course our own.

https://doi.org/10.1515/9783110986600-001

group.[1] Beria in general and the Wagi dialect[2] more specifically can be regarded as an underdescribed language. A grammar (Jakobi & Crass 2004), an unpublished BA thesis (Wolfe 2001) and a number of articles (e.g. Jakobi 2011, Wolfe & Abdalla 2018) all focus on the Kube dialect. For the Wagi dialect, on the other hand, first documentation efforts led to collections in two archives (Compes 2017, Compes & Hellwig 2022). Linguistic analyses, however, are still rare for this variety. Only an unpublished dissertation treating the basic areas of morphosyntax (Abdu El-Dawi 2010) and more recently an article on verbal morphology (Compes 2021) are available. From these treatments of Kube and Wagi, the following typological features of morphosyntax emerge: Beria's morphology is largely agglutinating. It has no case, but a rich verbal morphology. This verbal morphology includes a set of person indexes, derivational morphemes, and portmanteau morphemes for TAM, number and polarity. Its polypersonalism characterises Beria as a predominantly head-marking language on the clause level. It is verb-final, with APV as its basic, unmarked word order.

While this listing of grammatical features is largely based on segmental structures, Beria is also known to be a tone language. Importantly, tone plays a central role in the grammar, as it is the primary means to mark number distinctions both on verb forms and on nominals. The tone system, however, has not received much attention yet except for some preliminary remarks on verbal tone in Wolfe (2001) and some basic observations in Jakobi & Crass (2004). Although a much more thorough description of tone focusing on Wagi is underway as a PhD project by one of the authors, our knowledge of the tone system of any of the dialects is thus far from being satisfying at the present stage.

The starting point of our investigation of nominal number marking is thus the simple observation that in recordings of isolated words, the distinction of referent number is marked by a tonal contrast. Given the scarcity of previous research on the tone system of Beria and since no systematic description of tone patterns neither on nominals nor in general is available to serve as a basis, this investigation will be explorative to some extent. Therefore, especially the system of tone transcription and pattern interpretation adopted here (introduced and outlined in section 3.3) has a preliminary character and will remain on the floor of debate. By describing the marking of number on both isolated nominal forms and within noun phrases, however, we hope to also contribute to advancing knowledge on the nature and function of tonal markings in Wagi in general.

---

1 Estimations on numbers of speakers vary considerably: The *Ethnologue* (Lewis et al. 2015) reports 274.000 speakers in total. Osman 2006 reports around 180.000 speakers in Sudan.
2 In this paper, we will use the term Beria presenting general facts about the language while we will refer to Wagi, Kube etc. when talking about specific dialectal data and analyses.

The outline of the paper is as follows. In section 2, we will establish the background to our analysis of number marking. Next, in section 3, we will present the data, the methods of analysis and the theoretical framework applied in this investigation. The central part of the paper will then describe number marking on bare nominal forms in section 4 and in noun phrases in section 5. In conclusion, our findings will be discussed in section 6.

2 Background

We will start this section with general information on the phonology including the basic segmental transcription conventions (2.1), followed by some remarks on the current state of research regarding tone distinctions in Beria (2.2). The main part of this section is, however, devoted to nominals and the structure of the noun phrase in Wagi to provide the context for the analysis of nominal number marking. In section 2.3, we will thus introduce the lexical class of nominals and their morphosyntactic behaviour to prepare our treatment of nominals within a single system of tone classes in section 4. Moreover, the ordering of elements and the position of grammatical markers in the NP is shown, which later on can be compared with the locus of tonal number marking.

As already pointed out, the tone patterns of nominals and noun phrases are our research topic in sections 4–5. Tone patterns of other elements and structures such as verbs and clauses, however, have not been systematically investigated yet. With the exception of the presentation of basic tonal distinctions in section 2.2, we will thus not transcribe tone in the examples of the following subsections.

2.1 Phonology

Beria has a 7/9 vowel harmony system based on the feature Advanced Tongue Root [ATR]. Phonetically, there are nine vowels in Beria, four [-ATR] vowels, four [+ATR] vowels and a neutral mid vowel [ɑ] as depicted in Table 1 (cf. Jakobi & Crass 2004, Wolfe 2001).[3] However, as noted first by Anonby (2007) for Kube, the [+ATR] mid vowels [e] and [o] (put in parentheses in Table 1) can be analysed as allophones of the [-ATR] mid vowels /ɛ/ and /ɔ/ conditioned by one of the [+ATR] high vowels /i/ and/or /u/. Our data confirms this analysis resulting in a system with 7 vowel phonemes in Wagi.

---

3 Instead of IPA symbols with diacritics to represent the ATR-distinction, following other scholars in the context of African linguistics and more specifically Jakobi & Crass (2004: 19), we use IPA symbols corresponding to different vowel heights to encode the distinction.

Table 1: Wagi vowel inventory.

	[-ATR]		neutral	[+ATR]	
	front	back	mid	front	back
high	ɪ	ʊ		i	u
mid	ɛ	ɔ		(e)	(o)
low			ɑ		

In the current state of research, the domain for ATR vowel harmony is assumed to be the word. Vowels within this domain must agree on the feature [ATR] and the vowels of the two [+/-ATR] sets cannot be combined. The neutral mid vowel /ɑ/ can appear with either set. Affixes usually have two allomorphic variants, distributed according to the [ATR] value of the root vowels they attach to, as exemplified in (1).

(1) *gɪ-g-ɛ*   *gi-g-e*[4]
     open-A1sg-IPFV.sg   milk-A1sg-IPFV.sg
     'I open (sth)'   'I milk (e.g. a cow)'   Omda, p.c.

Alongside single vowels, Beria has the following rising and falling diphthongs: [ie, ei, ɛɪ, ɪɛ, ɪɑ, ɑɪ, ɑʊ, ɔɪ, ʊɪ, oi, ɔʊ, ou, ui, ue].

For consonants, Wagi exhibits an inventory of 16 phonemes, as represented in Table 2.

Table 2: Wagi consonant inventory.

	bilabial	alveolar	palatal	velar	glottal
plosive	b	t, d	ɟ	k, g	
nasal	m	n	ɲ	ŋ	
fricative		s			h
trill		r			
lateral		l			
approximant	w		j		

---

4 Examples display three to four tiers: (1) morpheme break of example, (2) tone pattern (optional) (3) interlinear glosses (optional), (4) engl. translation and a reference to the data source (see section 3.1). Furthermore, in the main body of the paper which presents the analysis of tone (sections 3-6), the first and second tiers are connected by so called association lines according to the conventions used in Autosegmental and Metrical Phonology (Goldsmith 1976, 1990, Hyman 2007). See section 3.3 and specifically FN 12 for an explanation of the tonal representation.

Alveolar and velar plosives distinguish between a voiceless and a voiced stop. The alveolar trill /r/ is usually realised as a tap [ɾ] intervocalically. The alveolar fricative /s/ has a postalveolar allophone [ʃ], both exist in free variation. Additionally, a voiceless bilabial fricative [f] can appear as an allophone of /b/ and also – before [-ATR] back vowels – as an allophone of /h/.

Furthermore, geminates are frequent in Wagi. All consonants except the approximants, /h/ and possibly some nasals exhibit long forms. So far, the status of this gemination is not clear. Geminates occur as a rule in some verb forms. In other cases, they are a feature of fluent speech. In a number of items (e.g. *ab.ba* 'father', *kad.da* 'good'), however, their origin is unknown and may well be lexical.

Beria does not allow for consonant clusters within one syllable (neither onset nor coda clusters). However, it does allow for open as well as closed syllables. The following syllable structures are attested: V, VV, CV, CVV, VC, CVC, and CVVC (as in ɔ 'milk', *oi* 'hair', *ba* 'hand', *bau* 'woman', *ur* 'belly', *bur* 'child', *tɪɛr* 'ax'). Roots are mostly mono- or disyllabic, sometimes (but rarely) trisyllabic. Longer words exist, but are usually morphologically complex.

Transcriptions throughout this paper will be phonemic using the symbols given in Tables 1 and 2. In line with speaker intuitions, the following allophones will, however, be represented by the respective phonetic symbol: [o], [e], [ʃ] and [f]. Geminates will be indicated by a doubling of the respective symbol (e.g. /dd/).

2.2 Tone

Beria is a tone language in the sense that it employs phonemic pitch distinctions. These pitch distinctions can be lexically contrastive as in (2), or mark grammatical contrasts, as e.g. number distinctions in verbs (see 3) and nouns (see 4).

Tonal minimal pair (lexical)

(2) *ba*   *ba*
ML[5]   MH
paternal.uncle.sg   hand.sg
'paternal uncle'   'hand'   ZAG_EOI_20150205_1

---

5 The representation of tone patterns adopted in this paper will be discussed in section 3.3. In the accent notation system typically used in the Africanist tradition, example (2) corresponds to a representation as *bâ* vs. *bǎ*.

Tonal minimal pair (grammatical: plural participants)

(3) *ja-r-ɛ*   *ja-r-ɛ*
    ML   MH
    drink-3A-IPFV.sg   drink-3A-IPFV.pl
    'he drinks'   'they drink'   ZAG_MAM_20181213_1

Tonal minimal pair (grammatical: referent number)

(4) *bʊr*   *bʊr*
    M   ML
    child.sg   child.pl
    'child'   'children'   Omda, p.c.

Three tone levels are currently assumed for Wagi: L, M, and H. Additionally, tone contours occur in all combinations of these three levels (i.e. LM, LH, MH, HM, HL, ML) and even complex contours seem to be possible, at least in the form of a rise-fall associated to one syllable as in (5).[6]

Complex tone contour

(5) *ta*
    LML
    head.pl
    'heads'   ZAG_EOI_20150205_1

2.3 Nominals and the structure of the noun phrase

Nouns in Beria are lexically distinguished from verbs in that they show very limited morphology compared to the rich morphology of verbs. The only morphological distinction on nouns, marked by a tonal contrast, is the plurality of the referent, which will be treated in detail in section 4. Syntactically, nouns can form an NP without any marking as in (6), where they constitute the core arguments, i.e. Beria has no core case markers. In predicative uses nouns are marked by a copula as in (7).

---

6 For Kube, Jakobi & Crass (2004: 32-33) propose three level tones and only two contour tones (LH and HL). Moreover, they suggest that the mid tone might not be underlying.

(6) *ɲa   oru   Ø-ʃɛ-rɛ*
    child.sg watermelon.sg P3-eat-A3-IPFV.sg
    'the/a child eats (a) watermelon'   ZAG_EOI_20141119_3

(7) *hɛrɪ   tarfʊ=i*
    falcon.sg bird.sg=COP.3sg
    'a falcon is a bird'   ZAG_EOI_21041112_1

In oblique and adjunct phrases, the noun phrase is marked by the clitics *=rI* 'LOC/ DIR' and *=rE* 'ABL'[7] as in (8) and (9).

(8) *bɪɛ=rɪ   ketti*
    house.sg=LOC   come.A3.PFV.sg
    'he came to the house' (also: *bɪɛr ketti*)   ZAG_EOI_20141119_1

(9) *bɪɛ=rɛ   ketti*
    house.sg=ABL   come.A3.PFV.sg
    'he came from the house'   ZAG_EOI_20141119_1

Further markers in certain focus constructions are the clitics *=gU* and *=dI*, whose functions are still the subject of debate (see Jakobi & Crass (2004: 151–154) for an analysis as focus markers and Wolfe & Abdalla (2018) for an analysis as an optional ergative marker and a copula).

Clitic markers of both of these sets (oblique/adjunct and focus constructions) attach to whatever is the last element of the hosting NP. This phrase-final marking appears only once per NP as in (10).

(10) *baʊ   lɪ=rɪ*
     woman.sg   QUANT=LOC
     'to a woman'   ZAG_MAM_20181129_3

Other than nouns, further nominals are personal and possessive pronouns, numerals, quantifiers, demonstrative pronouns, and a group of grammatical markers subsumed under the term *specifiers* here. Moreover, the open class of lexical items designating property concepts is a nominal class as well. All these elements can function as heads as well as modifiers in noun phrases. In the following, we will

---

7 Vowels in affixes undergoing [ATR] vowel harmony are represented by the respective capital letter throughout this paper.

focus on the modifiers selected for our investigation of number marking (see section 3.1 on the data set): property concepts, possessive pronouns and specifiers.

Lexical items designating property concepts (e.g. *ter* 'white' in (11), *bɪrɪ* 'brown' and *betti* 'big' in (12) and *kadda* 'good' in (13)) are almost identical to nouns in their morphosyntactic behaviour (tonal marking of referent number, unmarked in the function of core constituents, marked with the copula in predicative use, marked by clitics in oblique/adjunct phrases). They can thus be treated as a subgroup of the larger lexical category of nouns.

(11) *egi   tɛrr=i*
     POSS.1sg   white.sg=COP.3sg
     'mine is white'   Omda, p.c.

(12) *di   bɪrɪ   betti   ki*
     camel.sg   brown.sg   big.sg   DEM.prox
     'this big brown camel'   ZAG_EOI_20151214_1

(13) *ba   egi   kadda*
     paternal.uncle.sg   POSS.1sg   good.sg
     'my good paternal uncle'   ZAG_EOI_20150205_1

A regular use of these items, however, is in modifier function as illustrated in (12) and (13).[8] Since we investigated them exactly in this function, we will, for reasons of simplicity, refer to them as adjectives in the remainder of this paper.

The distinction between nouns and possessive pronouns, on the other hand, can be motivated insofar as the latter is a closed class forming a paradigm. Since part of the oppositions in possessive pronouns are marked by a tonal contrast, the paradigm will be treated in section 4.2 (see Table 5). Comparable to adjectives in their nominal behaviour, possessive pronouns can either appear as dependent elements in NPs (see 14), as independent pronouns forming an NP (see 11) or function as the complement of the copula (see 15).

(14) *bʊr   egi*
     child.sg   POSS.1sg
     'my child'   ZAG_EOI_20151208_3

---

8 In this modifier function nouns can be used as well, though.

(15) *bɪɛ=dɔ*   *egi=i*
house.sg=ANA₁ POSS.1sg=COP.3sg
'that house is mine'   ZAG_EOI_20160126_1

The small closed class of four operators which we call specifiers here – *ki* 'DEM. prox', *tɔ* 'DEM.dist', *=gI* 'ANA₁' and *=dO* 'ANA₂' (cf. Table 6 in section 4.2 for the whole paradigm) – can be analysed as a paradigm in the sense that, first, they are mutually exclusive in a noun phrase and cannot be combined, and second, they fill the same slot after all lexical items (cf. (12) above), but before other grammatical markers as in (16).

(16) *bɛttɪ*   *tɔ=rɪ*
tree.sg DEM.dist=LOC
'to that tree'   ZAG_MAM_20181129_3

Although the analysis of their function is still at the beginning, all four seem to encode the following meaning components: identifiability of the referent, presence of the referent in the discourse situation (i.e. *ki* and *tɔ* 'physically present', *=gI* and *=dO* 'mentally accessible'), and deixis (+/- proximity to deictic centre (DC)). Formally, the group is heterogenous. The specifiers *=gI* and *=dO* cliticize to the last lexical element of an NP as in (17). The specifiers *ki* and *tɔ*, however, are free forms (see 18) and can also be used pronominally as in (19).

(17) *di*   *egi=do*
camel.sg POSS.1sg=ANA₁
'my camel (that we talked about before)'   ZAG_EOI_20151214_1

(18) *hiri*   *ki*
cow.sg DEM.prox
'this cow'   ZAG_EOI_20160120_1

(19) *ki*   *ha=i*
DEM.prox stone.sg=COP.3sg
'this is a stone'   ZAG_EOI_20160120_1

Pronominal use is explicitly rejected for *=gI* and *=dO* by the informants. We thus differentiate between demonstrative specifiers (DEM.prox and DEM.dist) and anaphoric specifiers (ANA₁ and ANA₂).

The internal structure of noun phrases shows a head-initial order. There is, however, one exception to that rule: possessors encoded by phrases with full nouns precede the head as in (20), while pronominal possessors follow the head (compare (14) above).[9]

(20) *baʊ*   *bɔrʊ*
     woman.sg   man.sg
     'the man of the woman/a man of a woman'   ZAG_EOI_20141113_1

Nouns, adjectives, possessive pronouns and specifiers are combined in simple juxtaposition as in (12) above. If there is more than one modifying or determining element, possessive pronouns are always closest to the head (see (13) and (17) above), followed by other elements. Specifiers appear in phrase-final position. Thus, the structural template of a noun phrase in general is as follows:

HEAD (POSS) (ADJ) (SPEC)

To summarise the syntactic behaviour of the lexical items discussed: Nouns, adjectives, possessive pronouns and demonstrative specifiers behave almost identical. All can form an NP and generally remain unmarked when coding arguments. In predicative uses they are marked by the copula. Anaphoric specifiers, on the other hand, can only appear as modifiers. Depending on the semantics to be conveyed, NPs will be marked by clitics (e.g. =*rI* 'LOC/DIR' for a locative adjunct). This grammatical marking of NPs appears only once at the right edge.

3 Data and methods

This section will outline the methods of analysis and the theoretical framework used for our tone description. We will start with introducing our data (3.1), continue with the procedure of analysis of tonal patterns (3.2) and finally turn to the phonological interpretation underlying the system of tone transcription adopted here (3.3).

---

9 For Kube, Jakobi & Crass (2004: 140ff) report a second encoding strategy for possessor phrases with full nouns in which the possessor phrase is marked by *kii,* and follows the phrase coding the possessed. However, to date, this construction is not attested in our Wagi data.

3.1 Data

Our analyses to be outlined in sections 4 and 5 are primarily based on data compiled between 2014 and 2020 in a number of fieldwork classes led by Birgit Hellwig and Isabel Compes at the University of Cologne. This corpus is available online at the LAC archive (Compes & Hellwig 2022).[10] Tone annotations in the original ELAN transcripts were checked and mostly revised, as they represented the transcribers' initial hypotheses on the tonal system of Wagi and generally assumed a reduced tonal inventory compared to the present state of research. From this corpus, data of one speaker (EOI) was used. Additionally, both the analysis of tone classes in section 4 and the phonology sketch of the previous section (2.1) used data collected, annotated and analysed by Omda within his PhD project. This data set was recorded between 2019 and 2021 in fieldwork in Khartoum with a different speaker (ARJ).[11] Our investigation is thus based on data from two speakers, both male.

All recordings consist of elicitations from which we selected linguistic items (nouns, modifiers, and noun phrases) in contrastive singular-plural pairs for analysis. To ensure comparability of linguistic context, we limited our investigation to items uttered in isolation, since most of the archived recordings did not use a sentence frame. The items investigated can thus be regarded as all elicited within a "zero sentence frame" (Kutsch Lojenga 2018: 83). In this way, we could also make sure that no other, yet unknown, functions of grammatical tone in Wagi cloud the picture on tonal number marking. All repetitions of the target items available in the recordings (between 3 and up to around 8 per item, one item with only one repetition) were compared to abstract away from concrete phonetic realisations and identify the functional and categorical pitch events.

Since no previous research on the rather complex tone system of Wagi and especially on tonal interactions between elements in a phrase is available, we additionally limited our investigation of noun phrases to include three different structures only, all of them two-element phrases of one head noun and one modifier. The choice of specifically these three structures over other possible combinations was

---

10 The data for the analysis of NPs was taken from the following recording sessions: ZAG_EOI_20141113_1, ZAG_EOI_20150122_1, ZAG_EOI_20150205_1, ZAG_EOI_20151208_3, ZAG_EOI_20151214_1, ZAG_EOI_20160120_1, ZAG_EOI_20160126_1, ZAG_EOI_20160126_2. The interviewers in these sessions who also provided the first annotations in ELAN were Jan Junglas, KM, Melanie Schippling and Paul Compensis. Since the data on isolated nouns and other nominal forms are scattered over different recordings in the corpus, we will not list these recordings here. The reader is referred to the LAC archive instead, which provides respective metadata.
11 Data of a third speaker (MAM) available in the LAC corpus was excluded from our analysis, since considerable differences compared to the other speakers in our data seemed to exist in his grouping of words into tone classes.

largely governed by the availability of data in the corpus. Thus, section 5 is based on the analysis of two-word phrases of noun and possessive pronoun, noun and adjective, and noun and specifier.

3.2 Procedure and visual modelling of pitch events: *Periograms*

The data was analysed perceptually as well as by visual inspection of acoustic analyses in Praat (Boersma & Weenink 2017) and as *periograms* (Albert et al. 2018, 2019, Cangemi et al. 2019). Surface tone patterns of concrete utterances were then derived generalising over as many realisations of one phrase as available in the data by comparing both functional (singular-plural) and tonal minimal pairs. In this way, words and phrases could be grouped according to their tone patterns, resulting in seven distinct tone classes of bare forms. All groupings of words into tone classes have been checked with the native speaker intuitions of our co-author Omda.

For the visual inspection of acoustic analyses, we found periograms to offer an advantage over more widely used techniques of visualising plain F0 (fundamental frequency) trackings obtained, e.g. in Praat. Periograms are a recently developed technique of visually modelling pitch events from acoustic data by representing F0 trackings combined with measurements of what is called *periodic energy*. Periodic energy thereby refers to the strength of the periodic parts of the speech signal. As argued by Cangemi et al. (2019) and Albert et al. (2018), compared to a visualisation by F0 trackings alone, periograms thus give a representation of pitch events which is much closer to their actual perception, because periodic energy is directly relevant to the intelligibility of pitch (see Oxenham 2012).

We use such periograms for all examples that are visually modelled in sections 4 and 5. The visualisations were obtained using the ProPer workflow available on OSF (Albert et al. 2020), which includes a Praat script based on *mausmooth* (Cangemi & Albert 2016) and a number of R scripts (Rstudio: Rstudio Team 2019). Figure 1 exemplifies one of the periograms from our data.

In Figure 1, the upper broad curve traces F0 (y-axis) over time (x-axis), but, crucially, is also modulated (in thickness) for periodic energy: where periodic energy is strong and pitch thus easily perceptible, the curve shows broadly and clearly, where periodic energy gets weaker, the curve gets narrower and weaker as well. Where periodic energy is very low and pitch thus very hard to perceive, it vanishes completely. The model thus highlights perceptively relevant parts and enables a rather intuitive reading of the pitch events' visualisation. The periogram of *ta kadda* 'good heads' in Figure 1, for example, can be read as a realisation of rather level and low pitch on the first two syllables and a pitch rise clearly pronounced as a contour on the third and final syllable.

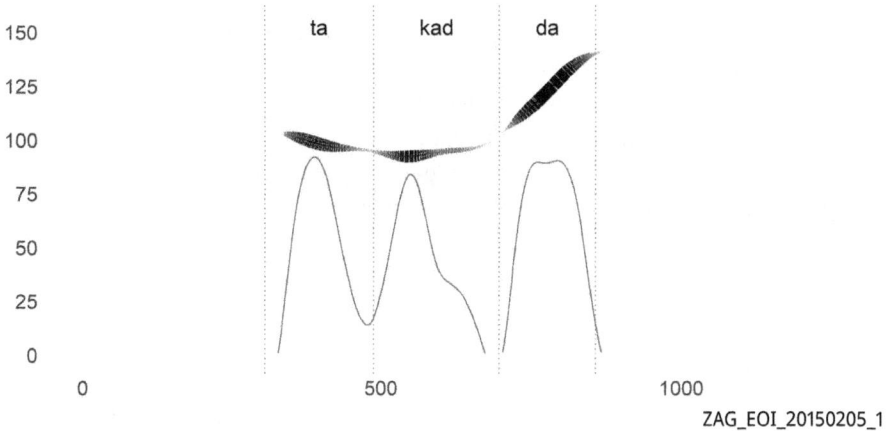

Figure 1: *ta kadda* 'good heads'; x-axis: time in ms; y-axis: F0 in Hz and periodic energy (thin line).

Additionally, the lower thin line in Figure 1 traces periodic energy (over time), with peaks in the line corresponding to vocalic segments (segments with high periodic energy) and troughs in the line to consonantals (segments with low periodic energy). The area between two troughs in this line can thus be roughly equalled to one syllable (see Cangemi et al. 2019 for a more detailed description). The dotted vertical lines, finally, represent syllable boundaries as manually annotated in Praat.

3.3 Theoretical framework and tone transcription: Word melodies

Although the phonological properties of the tone system in Wagi, and in Beria in general, have not been subject to much analysis and are in most parts not well established yet, we offer a phonological interpretation of our Wagi data here that will considerably facilitate our tone transcription and discussion of groupings of tone patterns and tonal processes (sections 4 and 5). Given the current state of research, however, this phonological interpretation, derived from our data on nouns and noun phrases, constitutes but a hypothesis on the phonological tone system of Wagi and serves as a descriptive tool. Its usefulness and appropriateness will be discussed in section 6.1 in light of the tone analyses presented in sections 4 and 5.

For this phonological interpretation, we draw on insights from the framework of Autosegmental and Metrical Phonology (Goldsmith 1976, 1990) and treat tones as autonomous segments, grammatically independent of the syllable or vowel they are associated with, representing them on a separate tier.[12] Apart from an association of equal numbers of tones and syllables, we thus assume that both an association of more than one tone with one syllable, as depicted in (21) for the example *ba* 'paternal uncle' and 'hand', and an association of one tone with more than one syllable, as given in (22) for the example *tɛlɛ* 'daughter', are possible. On the surface, a form like *ba* in (21) would result in a contour, a form like *tɛlɛ* in (22) in a level pitch over two syllables.

(21)   *ba*   *ba*
               ∧   ∧
              ML   MH
      'paternal.uncle.sg'   'hand.sg'   ZAG_EOI_20150205_1

(22)   *tɛlɛ*
               ∨
               M
          'daughter.sg'   ZAG_EOI_20160126_2

To enable generalisation over different syllabic word structures, we will furthermore treat tonal patterns as *word melodies* (see e.g. Gussenhoven (2004: 30ff)), i.e. as lexically (and grammatically) defined sequences of tonal targets distributed over the word according to its syllabic structure. The tone-to-syllable association will thus be explicitly indicated for examples where it is immediately relevant to the discussion only. For all other cases, it will be left unmarked, assuming a general association algorithm of tone sequences over words. This tone association is always a left-to-right association of one tonal target per syllable, with rightmost contouring where there are more tonal targets than syllables in a word. The word melody of a disyllabic word with a M tone on the first and a L tone on the second syllable, for example, will thus be indicated as ML, as will the word melody of a monosyllabic word with a ML falling contour. A word melody indication of LML, on the other hand, will indicate a L tone on the first and a ML falling contour on the second syllable for disyllabic words, but a LML complex contour for monosyllabic words. Tone

---

12 In the examples, associations of individual tones on the tonal tier with a specific segment on the segmental tier are represented according to the conventions of Autosegmental and Metrical Phonology, i.e. they are indicated by association lines. In the course of the paper, lines are also used to represent tonal processes (see section 5.2). In some of these tonal processes, tones are delinked from an associated segment in the lexical structure. This delinking is represented by a crossed-out association line: ǂ (see Hyman 2007, Kutsch Lojenga 2018 for identical representations of tone delinking in the context of tone spreading).

contours, under this association algorithm, are thus expected to occur on word-final syllables only. This assumption conforms to the situation previously observed in Beria (see Jakobi & Crass (2004: 33) for Kube). Such a left-to-right association with rightmost contouring is also a very likely pattern for word melodies cross-linguistically, known as the *association convention* since Goldsmith (1976, cf. Gussenhoven (2004: 32)).

Furthermore, word melodies are, in the cross-linguistically most widely attested case, by default not expected to exhibit sequences of like tones[13]. Instead, adjacent like tones in the surface structure are expected to result from spreading tones, which, in a word-melodic account, should be governed by a general spreading mechanism applying where there are less tonal targets than syllables in a word. Following the most widely attested pattern also coined into the association convention, we will assume a rightmost spreading mechanism here. Thus, a *M* melody on a disyllabic word will indicate a M tone spreading over both syllables of the word (as in (22) above), a *ML* melody on a trisyllabic word would result in a M tone on the first syllable and a L tone spreading over the second and third.

As hinted at above, the spreading behaviour of tones in Beria has, however, not systematically been investigated or established yet. While it fits most of our data, the spreading mechanism assumed here is thus rather to be taken as an initial hypothesis, implications of and problems with which will be discussed at the relevant points of data presentation. The usefulness of this hypothesis will be evaluated in section 6.1.

Moreover, our word-melodic account with tone patterns of individual tonal targets distributed over the word according to its syllabic structure entails a more general hypothesis on the tone system of Wagi: Assuming that phonological mechanisms generalise over different syllabic structures, i.e. that a ML melody on a disyllabic word is phonologically equivalent to a ML fall on a monosyllabic word, our analysis implies that all tone contours occurring in the surface form are phonologically derived, because they are simply a result of the left-to-right pattern of tone association. Although this is in line with the above-mentioned observations on Kube by Jakobi & Crass (2004: 33), the phonological status of contour tones, again, has not been clearly established yet, neither for Wagi nor for Beria more generally. The general usefulness and appropriateness of the word-melodic transcription as well as its implications will therefore be taken up as well in section 6.1.

---

13 This cross-linguistic dispreference for sequences of like tones is what is known as the Obligatory Contour Principle (see e.g. Gussenhoven (2004: 32f) for a short and concise discussion). Originally it was proposed by Leben 1980.

4 Tone classes and number marking on bare forms

4.1 Nouns

Nouns in Beria distinguish singular from plural forms by means of tonal marking, as visualised exemplary for the noun *kʊ* 'razorblade' in Figure 2.

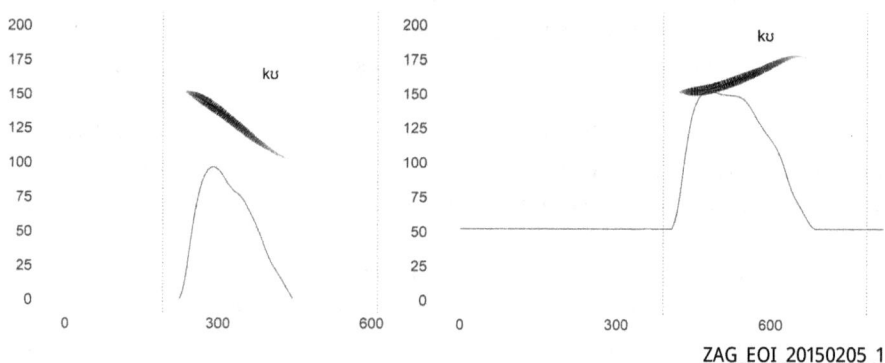

ZAG_EOI_20150205_1

Figure 2: *kʊ* ML 'razorblade' (left) vs. *kʊ* MH 'razorblades' (right).

For *kʊ*, the singular form is characterised by a pitch fall analysed as ML word melody, the plural form by a rising contour starting roughly at the same pitch level as the singular fall, which is thus analysed as MH word melody. For other nouns, however, the concrete tone patterns marking singular and plural forms respectively may be different, enabling the grouping of nouns into tone classes of singular and plural word melodies. Seven different tone classes are attested in our data, as given in Table 3. Example visualisations of all seven tone classes can be found in the appendix (Figures 5 to 11).

Table 3: Nominal tone classes.

Tone class	Tone pattern sg	Tone pattern pl	Example words
Class 1	L[14]	LML	*dɛrɪ* 'feather', *ɔ* 'person'
Class 2	LM	LML	*ɲarɪ* 'okra', *ta* 'head'
Class 3	LML	LLH	*koli* 'cheek', *ha* 'stone'

---

[14] L is the surface pattern attested for tone class 1. As will be suggested in section 5.2, LM as an underlying form might, however, be an alternative analysis for this group of words.

Table 3 (continued)

Tone class	Tone pattern sg	Tone pattern pl	Example words
Class 4	LLH	LHL	ʃeri 'mountain range', bʊ 'stick'
Class 5	M	ML	tɛlɛ 'girl/daughter', bʊr 'child'
Class 6	ML	MH	ʊrʊ 'throat', kʊ 'razorblade'
Class 7	MH	MHL	kɛbɛ 'ear', ba 'hand'

All of the word-melodic tone classes listed here include both mono- and disyllabic words, with rightmost contouring where there are more tonal targets than syllables in a word. Disyllabic nouns thus often exhibit a tone contour on their second syllable (e.g. class 3: ML fall in singular, LH rise in plural on the second syllable), while monosyllabic nouns in many cases even come with complex tone contours (e.g. LML or LHL rise-fall for plural forms of classes 1 and 2 and 4 respectively). Generally, singular melodies consist of one, two or three tonal targets, while plural melodies are always at least bitonal. Also, plural melodies end mostly on a L tone and for two tone classes (3 and 6) on a H tone, but never on a M tone, while singular melodies can end on any tone. Furthermore, not every logically possible combination of the three tone levels in Wagi (L, M, H) is attested as a nominal word melody here. That is, word melodies always start with a L or a M and never with a H tone. Assuming thus a constraint against word-initial H tones, still, for bitonal melodies, LH is absent as well. This might simply be a gap in the data, since LH word melodies are attested on other forms in Wagi, as e.g. on verbs (see Figure 3 below). For melodies with three tonal targets, however, more combinations are unattested, especially in singular forms. Possible tone patterns in three-tone melodies might thus be more restricted.

Additionally, the LLH word melody assumed for plural forms of tone class 3 and singular forms of tone class 4 in Table 3 requires a comment: Since it includes two adjacent like tones, LL, this pattern is not actually expected under our word melody account as outlined in section 3.3. However, it is well motivated from our data, because LLH sequences form a tonal opposition with LH sequences occurring, e.g. on verbs. Acoustically, this opposition has at least for disyllabic words a rather systematic correlate in the timing of the H target of the respective sequences, which is reached much later in the relevant syllable of a LLH word (LH contour on final syllable) than in the relevant syllable of LH words (level H tone on final syllable). This is exemplified in Figure 3 for the disyllabic word pair tɔrrɛ 'they dance' (LH), and koli 'cheeks' (LLH).

Since this LLH pattern cannot be accounted for in terms of the left-to-right spreading mechanism assumed for other forms, we do indeed treat it as including a sequence of two adjacent like tones here and transcribe the melody as LLH. The melodic pattern and its status in the tone system, however, clearly need a more thorough investigation left to future research.

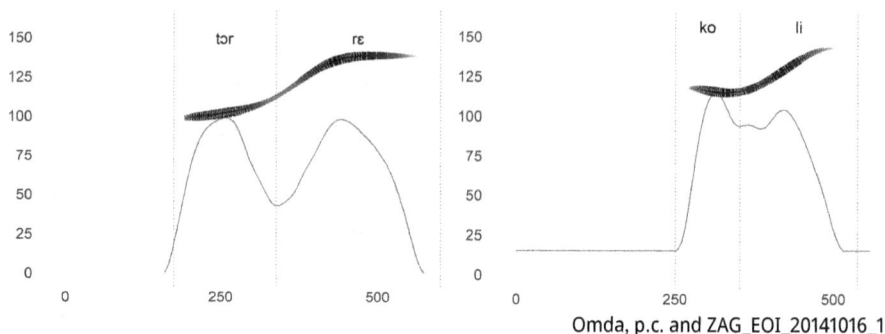

Omda, p.c. and ZAG_EOI_20141016_1

Figure 3: *tɔrrɛ* LH 'they dance' (left) vs. *koli* LLH 'cheeks' (right).

4.2 Other nominal forms

Apart from nouns, the same classification into tone classes can be applied to adjectives and possessive pronouns elicited as contrastive singular-plural pairs of isolated forms as well. That is, both these forms can be subsumed into the nominal tone classes given in Table 3. Four tone classes are attested for adjectives in our data, examples are given in Table 4.

Table 4: Attested tone patterns of adjectives.

Tone class	Tone pattern sg	Tone pattern pl	Example word
class 1	L	LML	*kʋa* 'untrue'
class 2	LM	LML	*ɟuse* 'tall'
class 3	LML	LLH	*kadda* 'good'
class 6	ML	MH	*tɛr* 'white'

Two adjectives, *betti* 'big, cl. 2' and *tetti* 'short, cl. 3', additionally mark their plural forms by a vowel contrast in our data, i.e. plural forms are *bettɛ* and *tɛttɛ* respectively.

Possessive pronouns belong to two tone classes: 1st and 2nd person pronouns to class 5 and 3rd person pronouns to class 6. Note that for possessive pronouns, tonal singular vs. plural refers to the number of the possessum, not of the possessor. That is, the possessive pronoun *kogu* '3pl.POSS; cl. 6' marked by a ML word melody (singular pattern), for example, indicates that they (3rd pl) possess something that is singular (cf. *tɛlɛ kogu* 'their daughter' in Figure 4 below). Number distinctions of possessors, on the other hand, are encoded segmentally in possessive pronouns (compare (23) to (24)). For clarification, the complete paradigm of possessive pronouns is given in Table 5.

(23) *tɛlɛ*   *kɔlɔ*
     M   M
     daughter.sg 2sg.POSS.sg
     'your (sg) daughter'   ZAG_EOI_20151208_3

(24) *tɛlɛ*   *kolu*
     M   M
     daughter.sg 2pl.POSS.sg
     'your (pl) daughter'   ZAG_EOI_20151208_3

Table 5: Singular and plural forms of possessive pronouns.

Person/Number	Segmental form	Tone pattern possessum.sg	Tone pattern possessum.pl	Tone class
1sg	*egi*	M	ML	class 5
2sg	*kɔlɔ*	M	ML	class 5
3sg	*kɔgɔ*	ML	MH	class 6
1pl excl	*karʊ*	M	ML	class 5
1pl incl	*karʊ*	ML	ML	no tonal opposition sg vs pl
2pl	*kolu*	M	ML	class 5
3pl	*kogu*	ML	MH	class 6

As becomes evident in Table 5, there is only one irregularity in this otherwise very systematic paradigm: 1st person plural pronouns tonally mark a distinction between exclusive and inclusive in singular forms. For 1st pl forms with a plural possessum, this clusivity distinction is, however, neutralised. Furthermore, the distinction between singular and plural possessums is neutralised as well for 1st pl.incl forms. This leads to a syncretism of three forms in this paradigm: 1st pl.incl possessum.sg, 1st pl.excl possessum.pl and 1st pl.incl possessum.pl (shaded cells in Table 5).

In contrast to possessive pronouns and adjectives, specifiers cannot be grouped into the seven tone classes of Table 3. The pattern for demonstrative specifiers attested in our data is a M melody for singular contrasting with a LH rise for plural (see Table 6). This pattern is not attested for nouns. Since the anaphoric specifiers =*gI* 'ANA₁' and =*dO* 'ANA₂' are not attested in isolation, we give their lexical melodies in brackets in Table 6. The assumption of a M (sg) vs LH (pl) melody is an analogy from the pattern attested for the demonstrative specifiers here. The analogy is justified by the tonal behaviour of the four specifiers within noun phrases, which is always identical.

Furthermore, the proximal specifiers *ki* and *=gI* employ a segmentally marked plural form (=)*sI*[15] combined with their tonal plural marking. The distal specifiers *tɔ* and *=dO*, however, mark number exclusively via tone (see Table 6). In this sense, the plural marking behaviour of specifiers can be regarded as differentiated according to their deictic component.

Table 6: Specifiers.

	sg	Tone pattern	pl	Tone pattern	sg	Tone pattern	pl	Tone pattern
	demonstrative				anaphoric			
proximal	*ki*	M	*sI*	LH	*gI*	(M)	*sI*	(LH)
distal	*tɔ*	M	*tɔ*	LH	*dO*	(M)	*dO*	(LH)

5 Number marking and the interaction of word melodies in two-word noun phrases

This section describes the tone patterns observable on noun phrases of two elements. Apart from an analysis of tonal number marking, it also gives a description of tonal processes resulting from the combination of two words within a phrase. The reason for this is as follows: For one, the tonal surface patterns on noun phrases in our data are the product of tonal number marking **and** additional tonal processes governing the interaction of word melodies within phrases. Second, and more importantly, we see the tonal processes observed as evidence for the phonologically derived nature of tone contours that we hypothesise in our word-melodic transcription system outlined in section 3.3. This derived nature of tone contours, in turn, will be a precondition for our account of a tonal plural morpheme to be discussed in section 6.2. The present section therefore has two objectives: the description of number marking in noun phrases (5.1) and the exposition of those tonal processes that occur systematically in our noun phrase data (5.2).

---

15 The demonstrative *sI* is a free form, while the anaphor *=sI* is a clitic that does not exhibit the expected vowel-harmonic allomorphy.

5.1 Number marking in noun phrases

In two-element noun phrases, number contrasts will be marked only once per NP, by a tonal opposition on the last element of the phrase, i.e. on the modifier. This is visualised exemplary for the noun phrase *tɛlɛ kogu* 'their daughter/s' in Figure 4.

The tone pattern of the contrast is that of the modifier's tone class, i.e. ML for singular and MH for plural realised on *kogu* '3pl.POSS, cl. 6' in Figure 4. Non-final elements in noun phrases, i.e. the head nouns, generally surface with the same tone pattern in singular and plural phrases (M for *tɛlɛ* 'daughter/s, cl. 5', see Figure 4). This non-number-contrastive tone pattern is identical with or derivable from the melody of the respective noun in its isolated singular form (M = singular melody of tone class 5). Phrase melodies of two-element noun phrases can thus schematically be described as singular melody of the noun followed by the singular or plural melody of the modifier according to phrase number (the number of the head noun's referent). Additionally, tonal processes apply to the combination of these two-word melodies. These will be discussed in the next section.

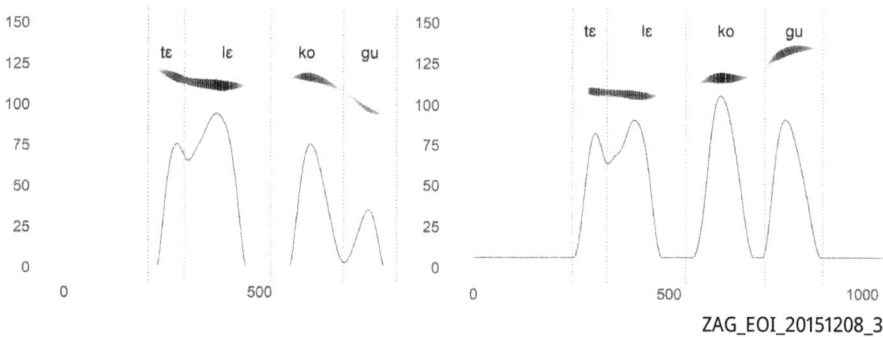

ZAG_EOI_20151208_3

Figure 4: *tɛlɛ kogu* M ML 'their daughter' (left) vs. *tɛlɛ kogu* M MH 'their daughters' (right).

5.2 Tonal processes in noun phrases

The combination of two words within one noun phrase seems to evoke various tonal processes applying to the relevant word melodies in Wagi. Three processes are attested systematically in our data and will be reported in this section:
1) H tone shift
2) L or M tone delinking
3) L vs. LM surface pattern in tone class 1

All of these processes occur in singular and plural noun phrases and apply to singular melodies of nouns followed by singular or plural melodies of modifiers.[16] Example visualisations of the three processes can be found in the appendix (Figures 12 to 14).

1) H tone shift
Where the (singular) word melody of a noun ends with a H tone (i.e. on nouns of class 7, MH, and class 4, LLH), in an NP, this H tone shifts to the syllable following the noun, i.e. to the first syllable of the modifier. This is depicted for the phrase *ba tɛr* 'white hand' in example (25).

H tone shifting from noun to modifier

(25) *ba*   *ba*   *tɛr*
       ∧   / ǂ ————— ǂ\
       MH   MH   ML
       hand.sg   hand.sg[17]   white.sg
       'hand'   'white hand'   ZAG_EOI_20150205_1, ZAG_EOI_20150122_1

As indicated in example (25), the relevant noun in these cases surfaces with just the first tone/s of its word melody, i.e. as M for nouns of class 7 (like *ba* 'hand') and as L(L) for nouns of class 4.[18] For disyllabic nouns of class 7, the M tone then spreads over both syllables of the noun, as depicted in example (26) for the phrase *kɛbɛ kɔlɔ* 'your (sg) ear'.

---

16 A number of smaller processes specific to certain elements or phrase types and not directly relevant to our discussion of number marking are, for reasons of clarity of presentation, omitted here. The interested reader is referred to the original MA thesis (Gayler 2021) for a complete listing of attested surface patterns.

17 In our examples of noun phrases, the gloss 'sg' or 'pl' indicates whether the (underlying) word melody of the respective item is the singular or plural melody of the item's tone class. It is not to be read as literally denoting the referent number of the noun phrase or a number morpheme on every item.

18 Thus, the H tone in the tonal sequence MH ML on *ba tɛr* in (25) is lexically associated to *ba* (crossed-out association line), but realised on the surface on *tɛr* (non-crossed-out association line). The second M tone of this same sequence is lexically associated to *tɛr* (crossed-out association line), but has no association in the surface structure (i.e. it is completely delinked). The last L tone of the sequence is realised on *tɛr* (non-crossed-out association line), where it is also lexically associated (no delinking). The surface melodic pattern on the sequence *ba tɛr* in (25) is thus M HL.

H tone shift from disyllabic noun

(26) *kɛbɛ*   *kɛbɛ*   *kɔlɔ*
       | | | /‡ ⎯⎯ ‡ /
       MH   MH   M
     ear.sg   ear.sg   2sg.POSS.sg
     'ear'   'your (sg) ear'   ZAG_EOI_20160126_1

Additionally, the shifting H tone delinks the first tonal association of the lexical melody on the target syllable (the first tone of the modifier's word melody), e.g. the M tone of the lexical ML melody on *ter* in example (25) or the M tone association with the first syllable of *kɔlɔ* in example (26). The surface melody of *ter* is thus HL instead of its lexical ML, that of *kɔlɔ* HM.[19]

Further conditions posed by the modifier's word melody, however, seem to block the H-shifting tonal process. Firstly, it cannot apply where the modifier's lexical melody starts with a L tone (modifiers of tone classes 1–4 and plural demonstratives). This means in fact that a shift occurs only where the first tone of the modifier to be delinked from the target syllable is a M tone (recall that no lexical melodies starting with a H tone are attested in our data). Where a L tone blocks the process, the noun-final H tone remains in situ, as exemplified for the phrase *ba kadda* 'good hand' in example (27). The tone class of *kadda* is class 3 (LML 'sg' vs. LLH 'pl').

Blocking: H tone in situ before L-initial word melody

(27) *ba*   *ba*   *kadda*
      ∧   ∧ | ∧
      MH   MH   LML
    hand.sg   hand.sg   good.sg
    'hand'   'good hand'   ZAG_EOI_20150205_1

Secondly, H tones cannot shift if the target tonal association to be delinked from the modifier is immediately followed by a H tone, i.e. if the shifting would create a sequence of two adjacent H tones. This is the case, e.g., in plural phrases with modifiers of tone class 6 (MH 'pl'). In contrast to a L tone as target of delinking, however, a target-adjacent H tone does not seem to block the process completely

---

19 If the delinked tone is the only tone on the modifier, as for example on singular specifiers, the shifted H tone will be the only surface tone on this modifier.

either: Where an additional H tone on the modifier is dispreferred because of the modifier's own H tone, noun-final H tones preceding these modifiers do not remain in situ, but are delinked completely instead. This is exemplified for the plural phrase *ba tɛr* 'white hands' in example (28). Series of adjacent H tones seem thus to be avoided in Wagi.

H tone delinking before modifier with a H tone as second tonal target

(28)
	ba		ba	tɛr
	∧		/⧧	∧
	MH		MH	MH
	hand.sg		hand.sg	white.pl
	'hand'		'white hands'	ZAG_EOI_20150205_1, ZAG_EOI_20150122_1

2) L or M tone delinking

Where bitonal word melodies of nouns do not end in a H tone (cl. 2: LM, cl. 6: ML), the second tone of these melodies is generally delinked when the noun is followed by a modifier. For disyllabic nouns, the first tone of the word melody then again spreads to cover both syllables of the noun. The phrase *dɪmɛ egi* 'my cat', for example, thus surfaces with the phrase melody L M, derived from the LM melody of *dɪmɛ* 'cat, cl. 2' and the M melody of *egi* '1sg.POSS, cl. 5', as depicted in example (29).

M tone delinking

(29)
	dɪmɛ		dɪmɛ	egi	
	| |		| /⧧	\ /	
	LM		LM	M	
	cat.sg		cat.sg	1sg.POSS.sg	
	'cat'		'my cat'		ZAG_EOI_20160126_1

Additionally, our data suggest that a similar delinking process might also apply to nouns of class 3 (LML 'sg'), surfacing with just a L tone in modified position. This is illustrated in (30) for the phrase *ha ki* 'this stone', which, however, is the only noun phrase relevant to the process that is attested in our data.

M and L tone delinking for cl. 3

(30) *ha*   *ha*   *ki*
     / | \   / ǂ ǂ |
    LML   LML   M
  stone.sg   stone.sg   DEM.prox.sg
  'stone'   'this stone'   ZAG_EOI_20151214_1

3) L vs. LM surface pattern in tone class 1

A final process to be observed with some regularity in our data applies to nouns of class 1, which generally surface with a LM instead of a L word melody within noun phrases. For example, the singular phrase *bɔrʊ kɔgɔ* 'his/her man' exhibits the surface melodic pattern LM ML (*bɔrʊ* 'man, cl. 1'; *kɔgɔ* '3sg.POSS, cl. 6'), as depicted in (31).

L to LM pattern change on nouns of cl. 1

(31) *bɔrʊ*   *bɔrʊ*   *kɔgɔ*
      \ / | ⤫ | | |
      L   LM   ML
   man.sg   man.sg   3sg.POSS.sg
   'man'   'his/her man'   ZAG_EOI_20160126_1

The same LM pattern can be observed as a contour on monosyllabic nouns of this tone class within NPs. This pattern change specific to tone class 1 might thus also allow for an interpretation of LM as the actual underlying word melody of this tone class with the M tone delinked in bare forms. Given the mechanisms of tone delinking depicted in 2), this interpretation seems even more likely than an account with underlying L melody and M tone insertion for modified forms, since tone delinking is attested elsewhere in the tone system, while tone insertion is not. Still, this interpretation also necessitates further research on why the delinking process applies to modified forms for tone classes 2, 3 and 6, but to bare forms for tone class 1.

To summarise, all three tonal processes described in this section are conditioned by certain tone patterns on nouns in initial position of two-word NPs and can thus be allocated to the tone classes with their respective patterns. Table 7 provides an overview over the tone classes and the respective process attested with each of them.

Table 7: Tone classes and attested processes in two-word NPs.

Tone class	Tone pattern sg	Example words	Tonal processes
Class 1	L or LM underlying?	*dɛrɪ* 'feather', *ɔ* 'person'	LM on modified nouns, L on bare forms
Class 2	LM	*ɲarɪ* 'okra', *ta* 'head'	M tone delinking
Class 3	LML	*koli* 'cheek', *ha* 'stone'	M and L tone delinking
Class 4	LLH	*ʃeri* 'mountain range', *bʊ* 'stick'	H tone shift
Class 5	M	*tɛlɛ* 'girl/daughter', *bʊr* 'child'	no process
Class 6	ML	*ʊrʊ* 'throat', *kʊ* 'razorblade'	L tone delinking
Class 7	MH	*kɛbɛ* 'ear', *ba* 'hand'	H tone shift

6 Discussion and conclusion

In discussing our findings, we start with some conclusions and remarks on the general tone system of Wagi, evaluating the hypotheses on its phonological properties incorporated into our transcription system outlined in section 3.3 (section 6.1). Building on the properties of the tonal system emerging in this discussion, we will then turn to forming conclusions on the marking of nominal number, both on bare forms and within noun phrases (section 6.2).

6.1 Phonological properties of the tone system in Wagi

For the tone system in general, most importantly, our outline of tonal processes in section 5.2 confirms that at least many of the tone contours in Wagi are indeed phonologically complex, as expected in our word-melodic analysis: their tonal targets engage in tonal processes individually. That is, the L tone of a ML contour (on one syllable) can be delinked while at the same time the M tone is not delinked, and similarly, the H tone of a MH contour (again on one syllable) can shift to the next syllable while the M tone does not shift, etc.

Furthermore, organising these individual tonal targets into word melodies as an abstraction over different word structures seemed to generally also be useful for the description here, since the tonal processes observed always made that abstraction as well: all tonal interactions described here pertained to words of all structures investigated (mono- and disyllabic words of different syllabic make-up).

For the assumed mechanism of spreading (left-to-right tone association with rightmost spreading within a word) and the expected accompanying dispreference of adjacent like tones, however, the picture is not as clear. On the one hand, there is evidence both for an active avoidance of adjacent like tones and for tonal spreading as such: The mechanism of H tone shift described in section 5.2 cannot fully apply if its outcome would be a series of two adjacent H tones, and both word-initial L and M tones can spread over two syllables within a word if the second tone of the respective word melody is delinked. On the other hand, the word-melodic pattern LLH, which is well motivated from the data, includes two adjacent L tones that cannot, unfortunately, be accounted for by our rightmost spreading mechanism. Both the initially hypothesised spreading pattern and the behaviour of the LLH word melody within the tone system thus need a more thorough investigation. Still, for the current purposes of describing the tonal marking of nominal number, the phonological interpretation of tone patterns as word melodies consisting of individual tonal targets seems to be quite useful.

As a final remark on the tone system, the processes described in section 5.2 suggest that H tones may have a special status in Wagi, not only because they are absent in initial position of lexical word melodies, but also because they appear to behave differently from the other tones in a similar context: both word-final L and M tones are delinked from modified nouns with bitonal word melodies; word-final H tones in the same position, however, are not delinked but retained, either in situ or shifted to the modifier (the only exceptions to this being motivated by the avoidance of HH sequences).

6.2 Nominal number marking

For the marking of nominal number, we conclude from our findings that it is achieved by a tonal plural morpheme attaching at the right edge of the noun phrase in Beria. This conclusion is based on the observation that number contrasts are marked only once per NP, by a tonal opposition realised on the phrase-final syllable. We suggest that this tonal opposition can be described as an opposition between a number-unmarked lexical word melody surfacing in singular forms and a combination of this lexical word melody with a tonal plural morpheme in plural-marked forms.

The reasoning behind this suggestion is as follows: as depicted in section 5, modified nouns appear with their singular melodies (or a pattern derived from their singular melodies and tonal interactions with adjacent elements) not only in singular but also in plural noun phrases (cf. *tɛlɛ* 'daughter' with its sg melody M in *tɛlɛ kogu* 'their daughters' in Figure 4 above). Singular melodies of nouns thus do not seem to explicitly indicate singularity of an NP's referent and are better described as number-neutral lexical word melodies. Furthermore, if tones within contours are treated as individual phonological entities as argued for in section 6.1, the plural tone patterns of several tone classes can straightforwardly be described as the tone classes' singular pattern plus one tone. This is the case for classes 2, 5 and 7, all with plural forms ending in a L tone. Assuming LM to be the actual underlying pattern of singular forms (cf. section 5.2), it would also apply to class 1. The added L tone can therefore be identified as the marker of plurality for these tone classes, the rest of the melody as lexically specified tone pattern.

For tone class 4, this description expects the surface melody LLHL as a plural pattern (LLH singular + L), which deviates from the attested form LHL. Again, this tone class with its LLH word melody needs further investigation and probably the assumption of an additional tonal process.

For tone classes 3 and 6, whose plural melodies end in a H tone, our tonal plural morpheme account requires the assumption of a H allomorph of the plural marker. We suggest that this allomorph appears where lexical melodies (i.e. the singular melodies of the relevant tone class) end in a L tone.[20] Still, both tone classes with H plural allomorph also require the assumption of additional tonal processes. This is in line at least with the observation that H tones in general seem to have a special status in the tone system of Wagi and behave differently in other contexts as well. The allomorphy of the tonal plural morpheme in Wagi is summarised in Table 8.

Table 8: Allomorphy of the tonal plural morpheme.

allomorph	context (lexical word melody)	tone classes	remarks
L	word-final M or H tone	2, 5, 7	
		1	(M tone delinked in bare forms)
		4	LLH+L> LHL
H	word-final L tone	3	LML+H> LLH
		6	ML+H> MH

The suggestion of a plural morpheme assigned once per NP at the right edge is in line with the behaviour of other grammatical marking in the noun phrase. For example, the oblique- and adjunct-marking clitics =rI 'LOC/DIR' and =rE 'ABL' described in section 2 attach once per NP at its right edge as well. The similar behaviour of tonal marking described in this paper highlights the generality of the process of phrasal marking in Beria. This phrasal marking covers a wide range of grammatical functions including plural marking, and it exhibits an equally wide range of formal means including a tonal morpheme.

---

20 For tone class 1, again, LM is assumed as the lexically underlying pattern.

Abbreviations

1	1st person
2	2nd person
3	3rd person
Ø	zero morpheme
A	agent
ABL	ablative
ANA	anaphoric specifier
COP	copula
DEM	demonstrative specifier
DIR	directional
dist	distal
excl	exclusive
H	High tone
IPFV	imperfective
incl	inclusive
L	Low tone
LOC	locative
M	Mid tone
P	patient
PFV	perfective
pl	plural
POSS	possessive pronoun
prox	proximal
QUANT	quantifier
sg	singular
TAM	tense-aspect-mood

Appendix

Example periograms of the seven tone classes (isolated forms)

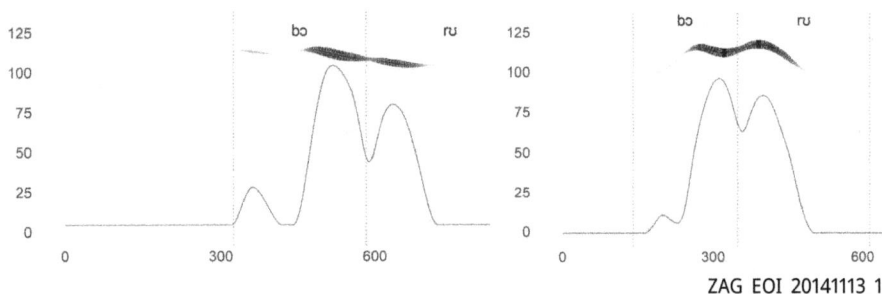

ZAG_EOI_20141113_1

Figure 5: Tone class 1: bɔrʊ L 'man' (left) vs. bɔrʊ LML 'men' (right).

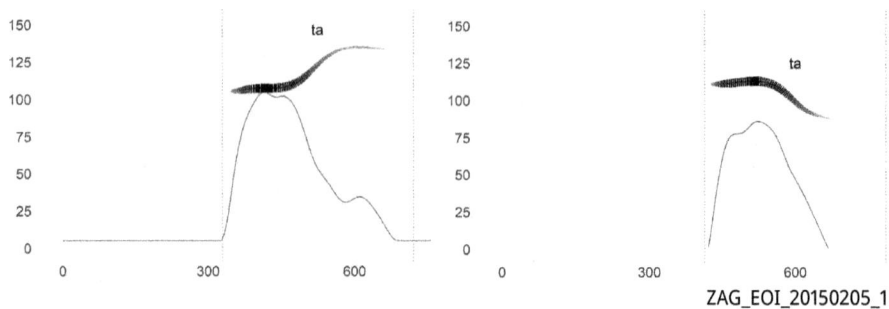

Figure 6: Tone class 2: *ta* LM 'head' (left) vs. *ta* LML 'heads' (right).

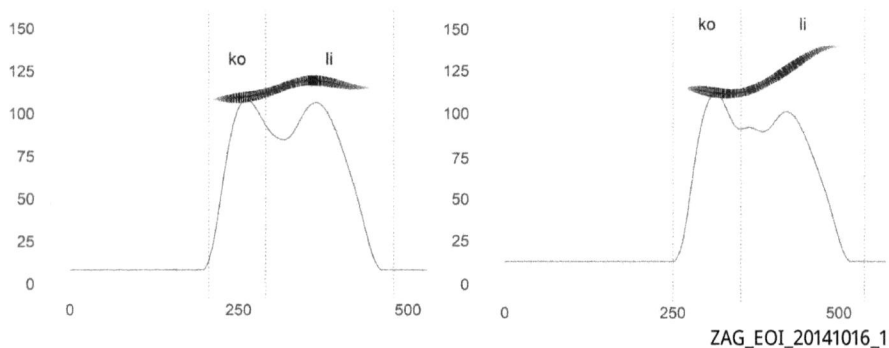

Figure 7: Tone class 3: *koli* LML 'cheek' (left) vs. *koli* LLH 'cheeks' (right).

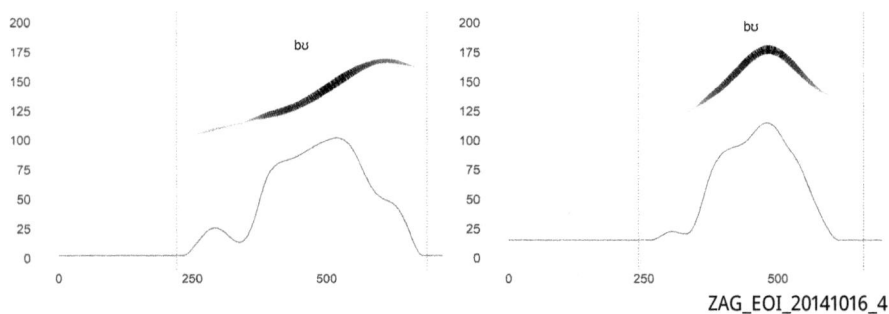

Figure 8: Tone class 4: *bʊ* LLH 'stick' (left) vs. *bʊ* LHL 'sticks' (right).

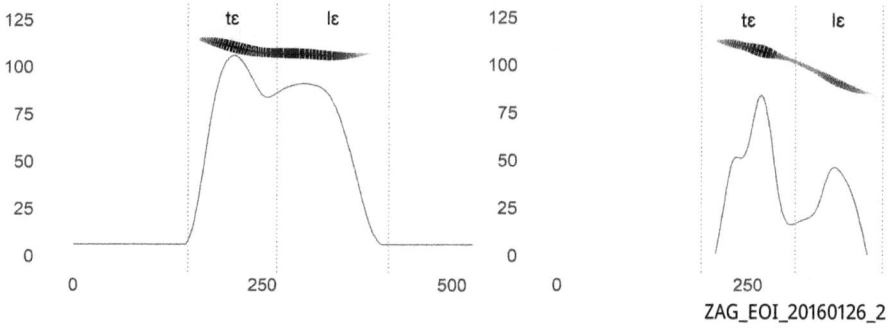

Figure 9: Tone class 5: *tɛlɛ* M 'daughter' (left) vs. *tɛlɛ* ML 'daughters' (right).

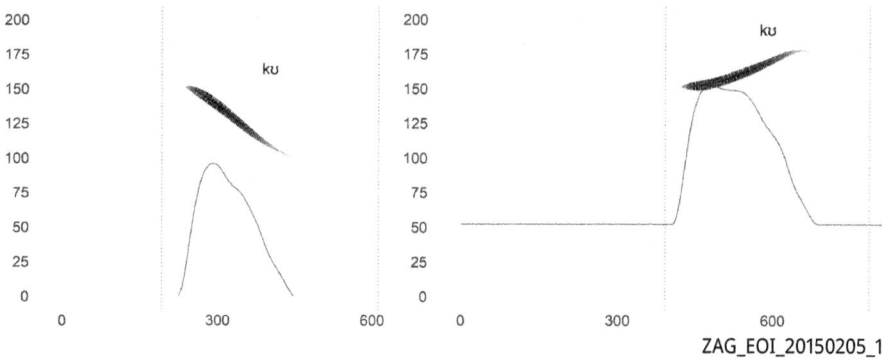

Figure 10: Tone class 6: *kʊ* ML 'razorblade' (left) vs. *kʊ* MH 'razorblades' (right).

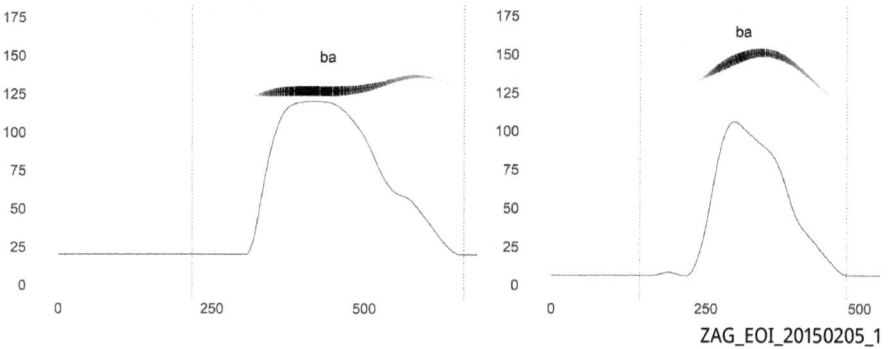

Figure 11: Tone class 7: *ba* MH 'hand' (left) vs. *ba* MHL 'hands' (right).

Example periograms of the three tonal processes (two-word NPs)

ad 1) H tone shift

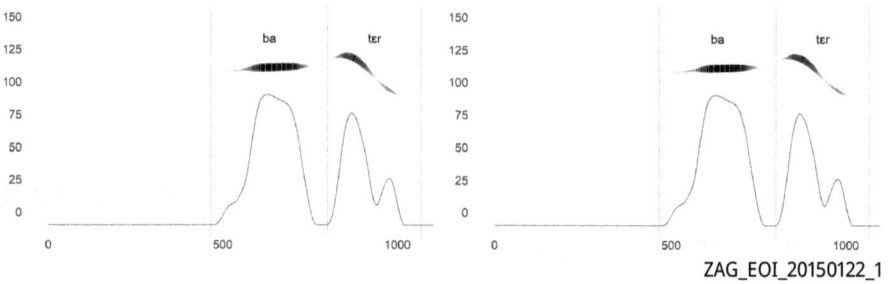

Figure 12: *ba tɛr* M HL 'white hand' < *ba* MH 'hand.sg' + *tɛr* ML 'white.sg'; H tone shifted from *ba* to *tɛr* (left) vs. *ba tɛr* M MH 'white hands' < *ba* MH 'hand.sg' + *tɛr* MH 'white.pl'; H tone delinked from *ba* (right).

ad 2) M tone delinking

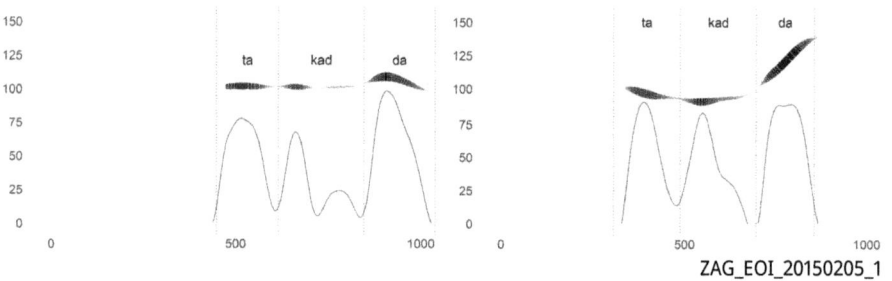

Figure 13: *ta kadda* L LML 'good head' < *ta* LM 'head.sg' + *kadda* LML 'good.sg' (left) vs. *ta kadda* L LLH 'good heads' < *ta* LM 'head.sg' + *kadda* LLH 'good.pl' (right); M tone delinked from *ta*.

ad 3) LM surface pattern in modified nouns of tone class 1

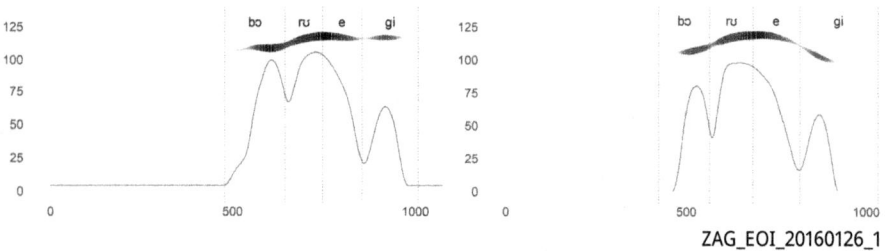

Figure 14: *bɔrʊ egi* LM M 'my man' (vs. *bɔrʊ* L 'man.sg' and *egi* M '1sg.POSS.sg' in isolation each) (left) vs. *bɔrʊ egi* LM ML 'my men' (vs. *bɔrʊ* L'man.sg' and *egi* ML '1sg.POSS.pl' in isolation each) (right).

References

Abdu El-Dawi Abdalla, Maha 2010. *The Morphosyntactic Structure of the Zaghawa language in Sudan, Focus: the Wegi dialect.* Khartoum: University of Khartoum dissertation.

Albert, Aviad, Francesco Cangemi & Martine Grice. 2018. Using periodic energy to enrich acoustic representations of pitch in speech: A demonstration. *Proc. 9th International Conference on Speech Prosody 2018*, 804–808. http://dx.doi.org/10.21437/SpeechProsody.2018-162

Albert, Aviad, Francesco Cangemi & Martine Grice. 2019. Can you draw me a question? Winning presentation at the *Prosody Visualization Challenge 2*. ICPhS, Melbourne, Australia. https://www.researchgate.net/publication/335096657_Can_you_draw_me_a_question?channel=doi&linkId=5d4e86644585153e5949fcb7&showFulltext=true

Albert, Aviad, Francesco Cangemi, T. Mark Ellison & Martine Grice. 2020. ProPer: PROsodic Analysis with PERiodic Energy. *OSF.* https://doi.org/10.17605/OSF.IO/28EA5 (accessed 15 March 2021)

Anonby, Erik John 2007. Book Review of Angelika Jakobi & Joachim Crass. 2004. Grammaire du beria (langue saharienne). *Journal of African Languages and Linguistics* 28(2). 217–220.

Anonby, Erik John & Eric Johnson. 2001. *A sociolinguistic survey of the Zaghawa (Beria) of Chad and Sudan.* Moursal-N'Djaména: Association SIL Tchad.

Boersma, Paul & David Weenink. 2017. *Praat: Doing Phonetics by Computer* (Version 6.0.33). [Computer program]. https://www.praat.org

Cangemi, Francesco & Aviad Albert. 2016. mausmooth: Eyeballing made easy. *Poster presentation at the 7th conference on Tone and Intonation in Europe (TIE).* Canterbury, UK.

Cangemi, Francesco, Aviad Albert & Martine Grice. 2019. Modelling intonation: Beyond segments and tonal targets. In Sasha Calhoun, Paola Escudero, Marija Tabain & Paul Warren (eds.), *Proceedings of the 19th International Congress of Phonetic Sciences, Melbourne, Australia 2019*, 572–576. Canberra, Australia: Australasian Speech Science and Technology Association Inc.

Compes, Isabel. 2017. *Zaghawa-Wagi: Towards documenting the Sudanese dialectal variant of Zaghawa.* Endangered Languages Archive. http://hdl.handle.net/2196/00-0000-0000-000F-BF52-A (accessed 31 August 2022)

Compes, Isabel 2021. The morphology of argument marking in Beria. *Studies in African Linguistics* 50(2). 196–226.

Compes, Isabel & Birgit Hellwig. 2022. *Zaghawa Fieldmethods Corpus 2014–2020.* Cologne: Data Center for the Humanities. Data deposited at the Language Archive Cologne (LAC). https://lac2.uni-koeln.de/de/zaghawa-wagi-dialect-sudan/ (site under construction).

ELAN (Version 5.3) [Computer software]. 2019. Nijmegen: Max Planck Institute for Psycholinguistics, The Language Archive. https://archive.mpi.nl/tla/elan

Gayler, Katharina. 2021. *Tone and Number in the Zaghawa Noun Phrase.* Cologne: University of Cologne MA thesis.

Goldsmith, John A. 1976. *Autosegmental Phonology.* PhD Dissertation, MIT.

Goldsmith, John A. 1990. *Autosegmental and Metrical Phonology.* Oxford: Blackwell.

Gussenhoven, Carlos. 2004. *The Phonology of Tone and Intonation.* Cambridge: Cambridge University Press.

Hyman, Larry M. 2007. Universals of tone rules: 30 years later. In Carlos Gussenhoven & Tomas Riad (eds.), *Tones and Tunes, Volume 1: Typological Studies in Word and Sentence Prosody*, 1–34. Berlin, Boston: Mouton de Gruyter.

Jakobi, Angelika. 2011. Split-S in Beria. In Doris Löhr, Eva Rothmaler & Georg Ziegelmeyer (eds.), *Kanuri Borno and Beyond. Current Studies on the Lake Chad Region*, 91–116. Cologne: Rüdiger Köppe Verlag.

Jakobi, Angelika & Joachim Crass. 2004. *Grammaire du beria (langue saharienne)*. Cologne: Rüdiger Köppe Verlag.

Kutsch Lojenga, Constance. 2018. Tone and Tonology in African Languages. In Augustine Agwuele & Adams Bodomo (eds.), *The Routledge Handbook of African Linguistics*, 72–92. London: Routledge.

Leben, William. 1980. *Suprasegmental Phonology*. New York: Garland.

Lewis, Paul, Gary Simons & Charles Fennig (eds.). 2015. *Ethnologue. Languages of Africa and Europe*. 18. Edition. Dallas: SIL.

Osman, Suleiman N. 2006. Phonology of the Zaghawa Language in Sudan. In Al-Amin Abu-Manga, Leoma Gilley & Anne Storch (eds.), *Insights into Nilo-Saharan Language, History and Culture*, 347–361. Cologne: Rüdiger Köppe Verlag.

Oxenham, Andrew. 2012. Pitch Perception. *The Journal of Neuroscience* 32(39). 13335–13338.

RStudio Team. 2019. *RStudio: Integrated Development for* R (Version 1.2.5019). [Computer software]. RStudio PBC: Boston, MA. http://www.rstudio.com/

Wolfe, Andrew. 2001. *Towards a Generative Phonology and Morphology of the Dialects of Beria*. Cambridge: Harvard University BA thesis.

Wolfe, Andrew & Adam Tajeldin Abdalla. 2018. Optional Ergativity and Information Structure in Beria. In Jason Kandybowicz, Travis Major, Harold Torrence & Philip T. Duncan (eds.), *African Linguistics on the Prairie: Selected Papers from the 45th Annual Conference on African Linguistics*, 341–358. Berlin: Language Science Press.

Deborah Arbes

2 Number inflection of English loanwords in Welsh

Abstract: The Siarad corpus offers an opportunity to investigate language contact between Welsh and English in spoken and informal language. The focus of this study lies on English multiplex nouns which occur in a Welsh context. Their integration into Welsh is measured on the basis of inflection for number and whether they undergo a soft mutation when expected. The findings reveal that insertions into the language, defined as single-word codeswitches, occur more often than English nouns with Welsh number inflection. Compared to Welsh and Latin nouns, English nouns integrated by Welsh number inflection show differences in their distribution across number categories and in the suffixes they employ.

Keywords: borrowing, codeswitching, loanwords, integration

1 Introduction

For several centuries the Welsh language has incorporated English loanwords into its lexicon. This is not surprising, considering the geographic, economic and political interlacing of England and Wales (see Section 1.2). In this paper, the focus will lie on nouns which have been borrowed from English while Latin nouns are also mentioned in passing. The word class of nouns promises to include more loanwords and borrowings than other parts of speech according to the borrowing hierarchies presented in Matras (2020). When investigating Welsh nouns, there is an opportunity to look into both the grammatical number and gender of loanwords. For this study, however, only number is examined as it is the more overt category. As Welsh and English have typological similarities (e.g. on a morphological level both employ suffixes), lexical and morphological borrowing is expected (see Thomason 2015).

Acknowledgements: To all Welsh speakers who have patiently answered my questions about plural forms, especially Iwan Rees, Sam Brown, Dafydd Weeks and the teachers at Cardiff School of Welsh: *Diolch o galon.* Also, I would like to thank Margaret Deuchar, Kevin Donnelly and Kevin Möllering for their help with accessing and sorting the corpus data and Jan Dyczmons for valuable advice on statistics. Furthermore, I am grateful to Thomas Stolz and Jago Williams for helpful comments and corrections. Despite having received lots of support, the sole responsibility for form and content of this article lies with me.

https://doi.org/10.1515/9783110986600-002

A Welsh street sign near Abergele involving several loanwords provides the first example:

(1) *Mae*   *'n*   *cymryd* *6 munud* *i*   *gerdded*
    be.3SG.PRES PTL take.VN 6 minute to ᴸwalk.VN
    at   *y*   **siopau,** **caffis** *a*   *mwy*
    toward   DET **shop:PL** **café:PL** and more
    'It takes six minutes to walk to the **shops, cafés** and more.'

The sentence in (1) shows two different strategies for pluralizing a foreign-language noun, namely the suffixes *-au* and *-s*. Adding a suffix is one of eight possibilities to mark the grammatical number of a Welsh noun. How often these suffixes occur and what other categories are relevant for English nouns will be answered in the following sections. A list of all multiplex[1] nouns mentioned in the conversations recorded for the Siarad corpus serves as a database for this project, in which the integration of borrowed nouns into Welsh is assessed through examining the means of number inflection. 'Lexical borrowings' and 'loanwords' are used as synonyms here and I follow Haspelmath (2009: 43) in using the terms "adaptation" and "integration" synonymously.

The following research questions concerning English nouns in a Welsh environment will be answered in this paper:

1a)   How frequently are English nouns pluralized by *-s* inserted into a Welsh environment?

1b)   Are these insertions integrated into Welsh by undergoing soft mutations?

2)   How are English loanwords distributed across the possible categories for number inflection?

Another relevant topic is the effect English morphology may have on Welsh nouns. It has been established that "the category of nominal plural has a higher-than-average borrowing rating" (Gardani 2015: 71). Examples for the plural suffix *-s* (sometimes realised as *-es* or *-ies*, represented as *-s* in the following) adapted by Welsh-origin nouns are mentioned by P. W. Thomas (1996: 175), e.g. *baswrs/baswyrs* (instead of *baswyr*) 'bass-players' and *pregethwrs/pregethwyrs* (instead of *pregethwyr*) 'preachers'. Alongside the noun *pilipalas* 'butterflies' these two forms are also found in the Patagonia corpus (see Arbes 2022). It is thus established that these forms exist and that borrowing of the plural suffix *-s* from the donor language

---

1 The term 'multiplex' refers to nouns denoting more than one entity and includes plural and collective nouns. The counter-part 'uniplex' is employed to pool singular and singulative nouns in one term (cf. Haspelmath & Karjus 2017).

English to the recipient language Welsh has taken place. The remaining research question is:

3) How frequently is the suffix -*s* borrowed and applied to Welsh-origin nouns in the Siarad corpus?

After answering this question there is an opportunity to compare the findings with written corpora and determine whether spoken and written language differ in terms of borrowing the suffix -*s*.

The remainder of this introduction is divided in two sub-sections. First, some historical background information about the interaction between Welsh and English is given and in 1.2, previous research on loanwords and closely related topics is discussed. Section 2 first introduces the corpus (2.1), describes the categories of Welsh number inflection (2.2) and draws a definitional line between loanwords and codeswitching (2.3). Following that, the frequency of single-word codeswitches is determined (3.1) before these nouns are analysed for possible soft mutations (3.2). Section 4 focuses entirely on loanwords with Welsh number inflection and zooms in on collective nouns (4.1) and borrowed nouns in category 1 (+suffix) (4.2). The last research question will be answered in Section 5, where the focus lies on Welsh-origin nouns pluralized by -*s*. Summaries of the results and concluding remarks are given in Section 6.

1.1 Historical background

English and Welsh have been in contact for centuries. Important dates in the history of English influence over the Welsh language are the year 1284, in which Wales was formally annexed to England, and the year 1536, in which English was declared the official language of Wales. Census data shows that in the beginning of the 20th century, Welsh speakers became a minority in Wales in that their numbers dropped below 50% of the population.[2] The second half of the 20th century was a period of improvements for the Welsh-speaking community, starting with the Welsh Language Act in 1967, which ensured everyone in Wales the right to use Welsh in legal proceedings. Another milestone of becoming and remaining a truly bilingual country was the introduction of Welsh as a compulsory subject for all pupils as part of the Education Reform Act 1988. In 2017 the Welsh Government set

---

2 https://www.visionofbritain.org.uk/census/EW1911WEL/3 (accessed 18.01.2022)

the ambitious goal of reaching one million speakers by 2050,[3] effectively doubling the number of Welsh speakers compared to the census in 2011. Receiving support for the language from the Government, and, in fact, even being represented by their own government, is a success for the community, which has seen Welsh become a minority language.

The census data shows a small number of Welsh monolinguals until 1981, but in the decades after that, monolinguals are not represented in the census, "as it was assumed that all respondents to the census would be able to speak English" (Deuchar et al. 2018: 7). Even if there are monolingual Welsh speakers nowadays (possibly under the age of three, before children are taught English in schools), Welsh-English bilingualism is the norm among Welsh-speakers and has been so for decades. The reality of bilingualism in Wales poses questions about social and linguistic aspects of the interaction between the two languages, some of which are addressed in this paper.

1.2 Previous research

Previous research related to bilingual communities includes studies on variation in dialects (e.g. Roberts 1988), obsolescence and revitalisation (Jones 1998) and Code-switching (e.g. Deuchar 2006, Deuchar & Davies 2009, Carter et al. 2011, Deuchar et al. 2016 and Deuchar et al. 2018). Thomas et al. (2014) and Binks (2017) focused their research on bilingual children and teenagers and their acquisition of grammatical features such as number and gender.

The integration of loanwords into Welsh has been examined for over a century. Early examples of studies concerned with loanwords are Lloyd-Jones (1910) for Latin loans into Welsh and Parry-Williams (1923) for English words in the Welsh language. However, Parina (2010) noticed that there was an imbalance between the amount of research conducted on Latin and English loanwords in Welsh, and that this might be for political reasons: "English loanwords are very often seen as marking a degraded stage of the Welsh language" (Parina 2010: 183). In order to provide an overview of the most frequently occurring loanwords in Welsh, she conducted a study using the text corpus *Cronfa Electroneg o Gymraeg* (CEG). The result is a list of 87 Latin borrowings and 40 English borrowings among the 1,000 most frequent words in CEG (Parina 2010: 185). This is a first step towards more detailed research including loanwords and borrowed nouns in the Welsh language.

---

3 https://gov.wales/sites/default/files/publications/2019-03/cymraeg-2050-a-million-welsh-speakers-annual-report-2017-18.pdf (accessed 18.01.2022)

The present paper expands on Parina's study and focusses on spoken varieties of Welsh rather than the written language.

In grammars and descriptions of Welsh, loanwords are usually discussed only very briefly. P. W. Thomas (1996: 162) mentions *-au* as the most common plural ending and the most likely one to be added to neologisms. King (2003: 51) reports that several suffixes are suitable to pluralize loanwords, firstly *-au* and *-ion*, (e.g. in *storïau* 'stories' and *egnïon* 'energies'). For nouns with a human reference, *-iaid* is a relevant suffix as it appears in loanwords such as *doctoriaid* 'doctors', *cwsmeriaid* 'customers' and *ffyliaid* 'fools'.[4]

There are, however, instances of nouns which are not fully integrated by a Welsh suffix but instead retain their English suffix. The plural forms ending in *-ys* are explained as retaining their English plural, "but in Welsh spelling" (King 2003: 62). Common examples for this are *bws – bysys* 'buses', and *nyrs – nyrsys* 'nurses'. P. W. Thomas (1996: 175) describes these forms as *"ffurfiau llafar anffurfiol"* ('informal spoken forms'). This is in accordance with A. Thomas (1987: 107), who declares frequent English loanwords as a way to distinguish colloquial from standard usage.

Previous studies are in disagreement over whether there is a clear difference between borrowing and codeswitching and where the line between them should be drawn. Matras (2020) places codeswitching and borrowing on a continuum and lists several characteristics for both ends. As all speakers recorded for the Siarad corpus are bilingual or multilingual, one of his definitions, i.e. that codeswitching mainly occurs between bilingual speakers while loanwords can be employed by monolinguals (Matras 2020: 117), is not applicable for this study.[5] Myers-Scotton (2002) uses frequency as a criterion to distinguish borrowed forms from code-switches: "a borrowed form will reoccur [. . .] because it has a status in the recipient language [. . .] The codeswitching form may or may not reoccur; it has no predictive value" (Myers-Scotton 2002: 41). Adopting this approach would result in several difficulties: firstly, the number of codeswitches would be dependent on the size of the corpus. A small corpus would hence produce a great number of codeswitches and vice versa. Secondly, highly morphologically integrated forms which may have been in use for a long time could be regarded as codeswitches if the subject they belong to was rarely mentioned and therefore the words might occur only once in the sources for the corpus. Rather than using frequency as a base for a definition, this study observes the frequency of codeswitches and compares it to the frequency of morphologically integrated loanwords (see 3.1).

---

4 Note that in the case of *ffŵl – ffyliaid* 'fool(s)' a vowel change applies as well.

5 One of his features by which to identify a borrowed noun, however, is applied in this study (see Section 2.2): "To treat borrowed nouns just like native nouns, and integrate them into native inflection patterns" (Matras 2020: 188)

Deuchar et al. (2018) reach the conclusion that codeswitching and borrowing are in fact two distinct phenomena. They suggest that "the defining characteristics of switches involve both low frequency and low levels of integration. Borrowings are different in that they may have low or high frequency, and will have high levels of integration" (Deuchar et al. 2018: 67). The study they refer to analyses English verbs which employ the Welsh verbal suffix -*io* (Stammers & Deuchar 2012). In addition to the use of the suffix, a feature called "mutation when expected" is observed as another way to assess integration into Welsh.

Initial mutations occur in all Celtic languages and cause some words to change their first consonant. Whether these mutations occur depends on the syntax of the sentence and the words preceding it (e.g. prepositions such as *am* 'about' or *i* 'to' call for a soft mutation in the following word). There are three types of mutations: soft, nasal and aspirate. Only soft mutations will be taken into account in this study, as they are the most frequently occurring ones (see Stammers & Deuchar 2012: 638). How a soft mutation affects a consonant is described in Table 1.

Table 1: Soft mutations in Welsh (cf. King 2003:14).

Original consonant (orthographic)	Original consonant (phonetic)	Soft mutation (orthographic)	Soft mutation (phonetic)	Examples		
p	[p]	b	[b]	*plant*	→ *blant*	'children'
t	[t]	d	[d]	*tegell*	→ *degell*	'kettle'
c	[k]	g	[g]	*cegin*	→ *gegin*	'kitchen'
b	[b]	f	[v]	*bara*	→ *fara*	'bread'
d	[d]	dd	[ð]	*defaid*	→ *ddefaid*	'sheep'
g	[g]	(disappears)	–	*gardd*	→ *ardd*	'garden'
ll	[ɬ]	l	[l]	*lloeau*	→ *loeau*	'calves'
m	[m]	f	[v]	*merch*	→ *ferch*	'girl'
rh	[r̥]	r	[r]	*rhosyn*	→ *rosyn*	'rose'

Having displayed the theoretical groundwork for this study, the next section will present the method employed in this paper including details about the Siarad corpus and the categories for Welsh number inflection.

2 Method

2.1 The Siarad corpus

Deuchar et al. (2018) describe the Siarad corpus as follows:

> Siarad (/ʃarad/) is the Welsh word for 'to speak' or 'speaking'. The *Siarad* corpus is a collection of 69 naturalistic recordings of conversations between bilingual speakers of Welsh and English. The recordings have been transcribed, glossed and translated, and can be found with links to the sound on <bangortalk.org.uk> and <talkbank.org>. The total corpus consists of about 450,000 words, or 40 hours, from 151 speakers. (Deuchar et al. 2018: 19)

Most recordings took place in or around Bangor (North West Wales) between 2005 and 2008, which is where 74% of speakers participating in the project grew up. 15% of participants were brought up in Mid or South Wales, 7% in North East Wales and 6% outside Wales (Deuchar et al. 2018). The participants "were invited to choose their own conversation partner and the place of recording" (Deuchar et al. 2018: 16) and the researcher was not in the room during the recording in order to ensure an informal setting for the conversations. After transcribing the conversations, the texts were glossed automatically by Bangor Autoglosser[6] and made available for download at http://bangortalk.org.uk/. The glosses I included in this paper are however are not identical to the output of the Autoglosser as I added some translations which the Autoglosser did not recognize and applied the Leipzig Glossing Rules.

2.2 Categories of Welsh number inflection

The way number inflection is categorized in this study is based on P. W. Thomas' (1996: 155) observation that two processes, namely addition and exchange of elements, play a role. The elements which can be added are plural suffixes or singulative suffixes. The elements which are exchanged can be suffixes or vowels. A combination of these processes is visualized in Table 2, completed by a suppletive category, in which the stem is exchanged. A similar model for explaining pluralization categories is put forth by Thomas et al. (2014).

---

6 http://bangortalk.org.uk/autoglosser.php (accessed 05.05.2021)

Table 2: Categories of Welsh number inflection.

No. of category	Description	Morphological explanation		Translation of example
		Uniplex form	Multiplex form	
1	+ suffix	base	base + suffix	
		peth	*peth-au*	'thing(s)'
2	+ suffix + sound change	base	base + suffix + sound change	
		gair	*geir-iau*	'word(s)'
3	~ suffix	base + suffix 1	base + suffix 2	
		hog-yn	*hog-iau*	'boy(s)'
4	~ suffix + sound change	base + sound change + suffix 1	base + suffix 2	
		deigr-yn	*dagr-au*	'tear(s)'
5	collective	base + suffix	base	
		coed-en	*coed*	'tree(s)'
6	collective + sound change	base + sound change + suffix	base	
		plent-yn	*plant*	'child(ren)'
7	vowel change	base	base + vowel change	
		car	*ceir*	'car(s)'
8	suppletive	base 1	base 2	
		person	*pobl*	'person/people'

2.3 Loanword or codeswitching?

This study regards those nouns as 'integrated' which employ a Welsh plural suffix or another type of Welsh number inflection in its uniplex and/or multiplex form (see Table 2). In contrast, the term 'codeswitching' is applied for nouns which

a)   are of English origin
     AND

b)   are inflected by the English plural suffix *-s* (or one of its allomorphs).

Nouns employing the plural suffix *-s* or *-ys* and which are additionally subject to Welsh number inflection, e.g. by undergoing a vowel change or employing a singulative suffix, are instead categorized as loanwords.[7]

A list of all multiplex nouns mentioned in the Siarad corpus was created. English nouns which are followed or preceded by another English word were deleted; this includes nouns from complete monolingual English sentences as in (2) as well as nouns which occur in so called Embedded Language Islands (ELI) (see Deuchar 2018, Myers-Scotton 2002) illustrated in (3).

(2) Lloyd-11, (855) GRG
     *Er she doesn't get paid during the **holidays.***

(3) Davies-15, (276) TEG

dw	*i*	*meddwl*	*bod*	*fi*	*mynd*	*am*	***few***	***days***
be.1SG.PRES	1SG	think.VN	be.VN	1SG	go.VN	for	**few**	**day:PL**

'I think I'm going for a **few days**'

Thus, all remaining English nouns are either single-word codeswitches as depicted in (4) or English loanwords (5), which employ Welsh number inflection.

(4) Fusser-05, (588) DYF

Mae	*nhw*	*yn*	*cyrraedd*	*mewn*	***boxes***	*o*	*dau*	*ddeg*
be.3SG.PRES	3PL	PTL	arrive.VN	in	**box:PL**	of	two	ᴸten

pump	*fel*	*arfer*
five	like	habit

'They arrive in boxes of 25, usually'

(5) Roberts-04, (991) LIL

O	*'n*	*i*	*licio*	*'r*	***cwestiynau***
be.1SG.IPFV	PTL	1SG	like.VN	DET	**question:PL**

'I liked the questions'

The etymology for each multiplex noun was determined with the help of Geiriadur Prifysgol Cymru (GPC). Subsequently, each noun was assigned to one of the eight categories of the Welsh number inflection system (see Table 2). Therefore, loan-

---

7 The terms "loanword" and "codeswitching" are chosen as a makeshift label in order to have a name for two concepts, whose differences are visible on the surface of the words. I do not intend to make theoretical implications by the labels. The same approach is taken in Arbes (2022).

words and codeswitches could be analysed separately. Their frequencies and the preferred categories for loanwords are presented in the following sections.

3 Codeswitching

Before getting into the analysis of single-word codeswitches in the Siarad corpus, some background information drawn from previous studies is given. Those studies focus on Codeswitching as a whole without excluding any parts of speech. Additionally, sociolinguistic variables are taken into account, while the present study focusses on only one part of speech (multiplex nouns). Some of the studies mentioned in 1.2 conducted by Deuchar and colleagues revolve around Codeswitching with regards to the MLF and the Matrix Language turnover hypothesis (Myers-Scotton 2002, 1998). One of the findings from Deuchar & Davies (2009) is that classic codeswitching, as found in the Siarad corpus, with Welsh as the Matrix Language providing the grammatical frame, is a sign of stable bilingualism. Extensive Codeswitching under these conditions is therefore not necessarily a reason to be concerned about language shift towards English. Furthermore, Deuchar et al. (2016) conclude that firstly, the use of bilingual clauses decreases as the age of speakers increases. This may be due to a more positive attitude towards codeswitching held by younger speakers and possibly indicates a long-term diachronic change towards more Codeswitching in Welsh. Secondly, they found that "those speakers who had acquired both Welsh and English from birth were significantly more likely to produce intraclausal code switching than all other categories of speakers" (Deuchar et al. 2016: 233), which suggests that a lack of fluency is not the reason why speakers use bilingual clauses.

3.1 Welsh number inflection vs. single-word codeswitches

Figure 1 provides an overview of all multiplex nouns in the Siarad corpus.

47% of types are English single-word codeswitches employing the plural suffix -*s*. Loanwords as well as Welsh-origin nouns employing Welsh number inflection account for 53% of types and 82% of tokens. Interestingly, it is possible for one lexeme to occur several times in this representation. An example is *chwarae-on* 'game-s' which occurs as a Welsh noun applying Welsh inflection (i.e. the suffix -*on*), as an English noun applying Welsh inflection (*gem-au* 'games') and as an English noun employing the suffix -*s*: *games*.

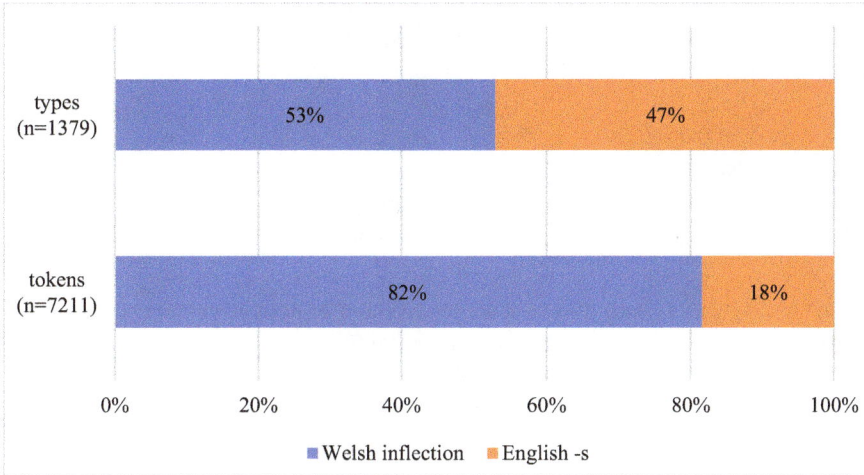

Figure 1: Welsh number inflection vs. single-word codeswitching.

Focusing on English-origin nouns, Figure 2 shows what percentage of English nouns adopts Welsh number inflection or retains the plural suffix -*s*.

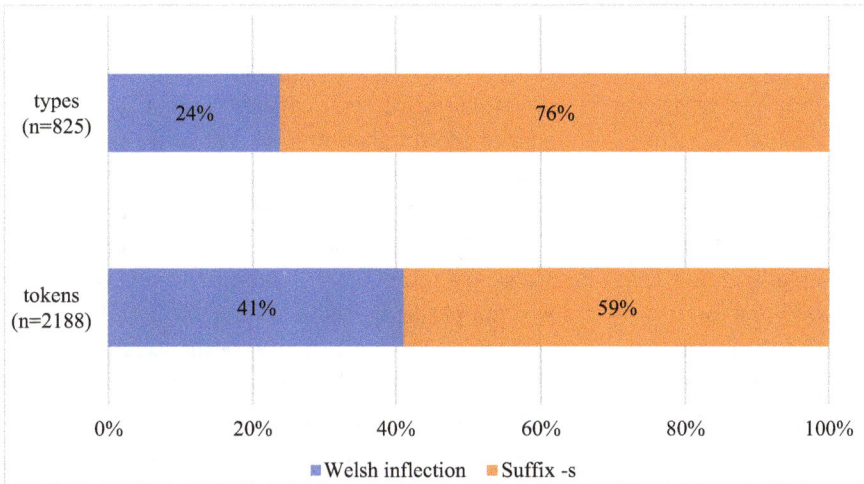

Figure 2: English nouns: Borrowing or Single-word codeswitching?

English-origin nouns are more likely to appear as codeswitches than to be integrated by Welsh number inflection. There are nouns which appear only as codeswitches and do not have an integrated loanword as an equivalent, e.g. *ages*. The choice speakers have in these cases is between the English form as in (6) or the Welsh form (*oesoedd* 'ages').

(6) Davies-12, (1058) CER
    A   *oedden*   *ni*   *allan*   *am*   ***ages***
    and   be.1PL.IPFV   1PL   out   for   **age:PL**
    'And we were out for **ages**'

However, the corpus also contains lexemes which may appear as codeswitches and as integrated loanwords, as illustrated in (7) and (8).

(7) Fusser-21, (1045) ILI
    ac   *oedd*   *o*   *yn*   *cario*   ***bagiau***
    and   be.3SG.IPFV   3SG.M   PTL   carry.VN   **bag:PL**
    'and he was carrying **bags**'

(8) Davies-12, (1316) SAL
    handio   *allan*   ***bags***   *yna*
    hand.VN   out   **bag:PL**   there
    'handing out those **bags**'

The type-token-ratio in Figure 2 shows that the integrated loanwords occur more frequently per type, whereas the single-word codeswitches show a tendency to occur less often per type. A considerable number, namely 396 single-word code-switches occur only once, amounting to 31% of nouns in this category. Compared to 8.6% hapaxes among integrated loanwords, the difference is significant.

This result supports Myers Scotton's (2002) approach to codeswitching only to some extent, stating that loanwords are more likely to reoccur in a corpus than codeswitching forms, as they have "no predictive value" (Myers-Scotton 2002: 41). In the present study, loanwords do reoccur at a higher rate than codeswitches, however the majority (69%) of what are defined as codeswitches also reoccur (between one and 31 times per type, *loads* 'loads' being the most frequent).

This high number of codeswitches and their frequent reoccurrences indicates that these forms have become an integral part of spoken Welsh. It also provides an answer to research question 1a).

Having found out that English single-word codeswitches are a very frequent phenomenon in the Siarad corpus, even if some of them are only mentioned once, it is worthwhile taking a look at how they may be integrated into Welsh by soft mutation.

3.2 Codeswitching: Mutated when expected?

An initial consonant mutation applied to an English-origin noun is evidence of a degree of integration into Welsh. The noun is not simply inserted in its original form, but an adaption to Welsh morpho-syntax has taken place. The following sub-section explores to what extent single-word codeswitches apply soft mutations and how the frequency of mutated codeswitches compares to that of loanwords and native Welsh nouns.

In order to determine what percentage of English plural nouns are mutated when expected in a Welsh context, a random sample consisting of 129 nouns (10% of all single-word codeswitches in the corpus) was created. For comparison, two more random samples (10% of loanwords and 10% of Welsh-origin nouns) were created, consisting of 67 and 320 nouns respectively.

As Figure 3 shows, only five out of 21 English plural nouns underwent a soft mutation when they were expected to.

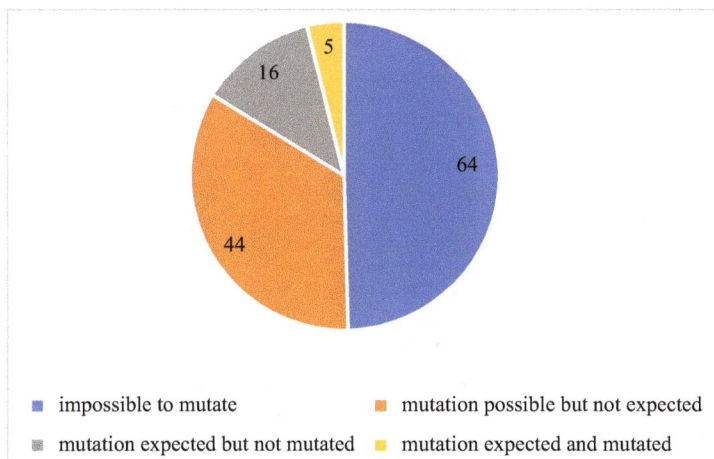

Figure 3: Random sample of single-word codeswitches (n=129).

Example (9) shows a case in which an English noun was mutated when expected (because of the preposition *o* 'of'), whereas example (10) illustrates the more common case that a single-word codeswitch is not integrated into Welsh by applying a soft mutation.

(9) Davies-01, (726) NON:
     couple o ***ddrinks*** *ar_ôl practice wedyn wastad gyda ni*
     couple of [L]**drink:PL** after practice afterwards always with 1PL
     'We always had a few **drinks** after practice then'

(10) Fusser-15, (1286) GFR:
      ges i ***panics*** *dydd Sadwrn*
      [L]**get.1SG.PST** 1SG **panic:PL** day Saturn
      'I got **panics** on Saturday'

5 out of 21 is a rather low ratio when compared to loanwords and native Welsh plural nouns.

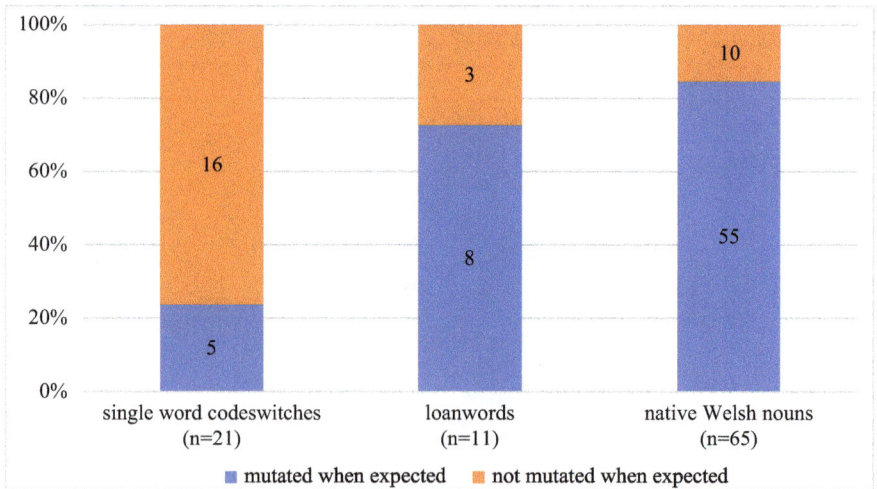

Figure 4: Mutation when expected.

Even though the data for loanwords should be taken with a grain of salt, as the sample was smaller than the other two, Figure 4 shows that loanwords and native Welsh plural nouns mutate at much higher rates. Example (11) shows a case of a loanword in which a soft mutation is triggered by *yna* 'there'.

(11) Fusser-18, (978) ARD:

a	*wedyn*	*mae*	*yna*	***gwestiynau***	*ia*	*am*
and	afterwards	be.3SG.PRES	there	**question:PL**	yes	about

y	*paragraph*	*bach*	*yna*
DET	paragraph	little	there

'and then there are **questions**, yes, about that little paragraph'

Figure 4 provides the answer to Question 1b) "Are single-word codeswitches integrated into Welsh by undergoing soft mutations?" Of course, not all single-word codeswitches behave the same. However, a random sample has shown that while mutated single-word codeswitches do occur, they are in a minority.

The above mentioned results are in consonance with findings from Stammers & Deuchar (2012), whose study offers "empirical evidence of a continuum of integration, from verbs that are not integrated to verbs that are fully integrated" (Stammers & Deuchar 2012: 642). The examples above depict a range from unmutated single-word codeswitches, which can be viewed as insertions into the language, via single-word codeswitches which are nevertheless mutated, to loanwords which mutate when expected and are therefore on the most integrated level of the continuum.

Having touched upon the integration of loanwords by soft mutation, the next section will take a closer look at how loanwords are integrated by Welsh number morphology.

4 English loanwords

This section revolves around those nouns which apply Welsh number inflection. The focus of this section lies on English loanwords from various phases (e.g. Old English, Middle English and Modern English), although Latin loanwords are mentioned as well. Figure 5 visualizes the percentages of the three origin languages Welsh, Latin and English.

Unsurprisingly, Welsh-origin nouns are in the majority (e.g. *blodau* 'flowers'). Regarding types, English nouns come second, however, regarding tokens, Latin nouns occur more frequently than English nouns. Frequently occurring Latin loanwords are e.g. *pobl* 'people' and *plant* 'children' (see Table 3).

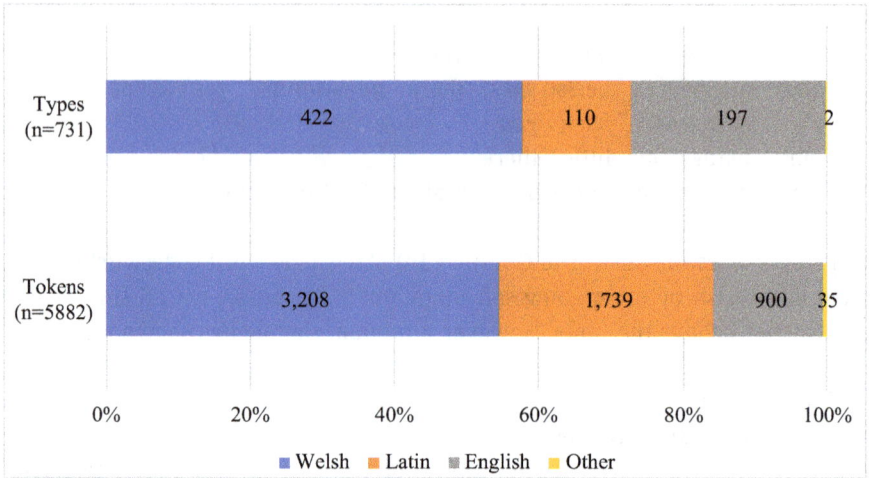

Figure 5: Origin Languages of nouns with Welsh number inflection.

Figures 6 and 7 illustrate the shares of the individual categories among Welsh nouns and borrowed nouns. The graphs for Welsh origin nouns serve as benchmarks with which to compare the borrowed nouns.

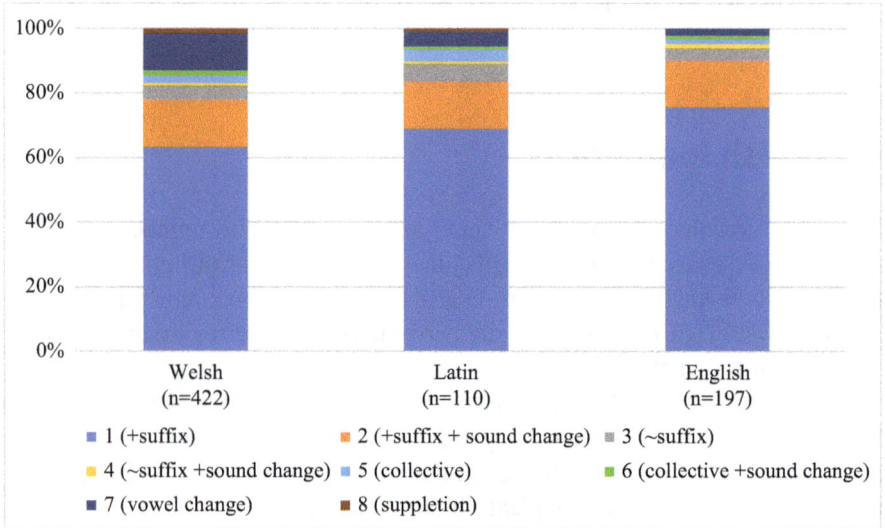

Figure 6: Nouns with Welsh number inflection (Types).

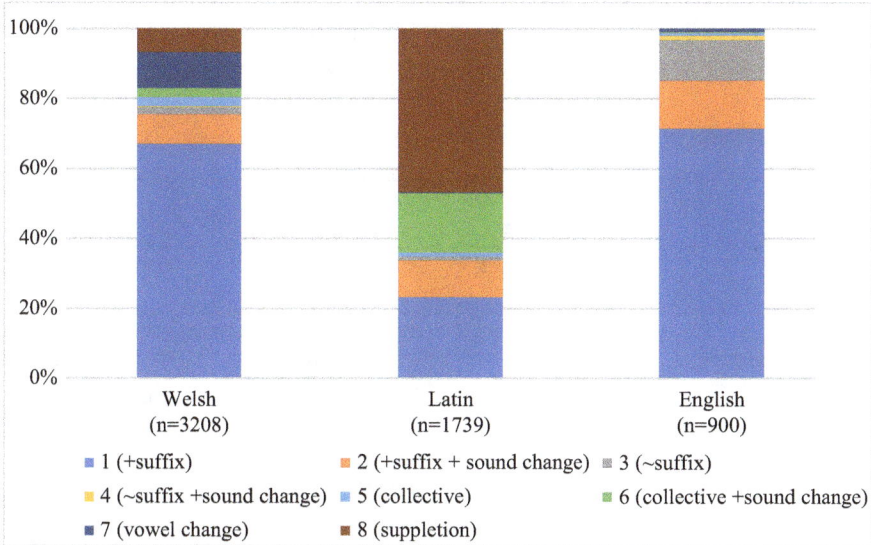

Figure 7: Nouns with Welsh number inflection (Tokens).

Category 1 (+suffix) is the most common way to pluralize Welsh origin nouns, followed by category 2 (+suffix +sound change). For Latin nouns, the distribution of types looks very similar (see Figure 6), however it is striking that some nouns in categories 6 and 8 occur disproportionally often. Table 3 shows some examples of frequent Latin loanwords next to Welsh-origin nouns of the same categories.

Table 3: Examples for Latin loanwords.

Category	Welsh			Latin		
Cat 1 (+suffix)	*dydd*	→	*dyddiau* 'day(s)'	*gwers*	→	*gwersi* 'lesson(s)'
Cat 2 (+suffix + sound change)	*gair*	→	*geiriau* 'word(s)'	*anifail*	→	*anifeiliaid* 'animal(s)'
Cat 6 (collective +sound change)	*adar*	→	*aderyn* 'bird(s)'	*plant*	→	*plentyn* 'child(ren)'
Cat 8 (Suppletive)	*ci*	→	*cŵn* 'dog(s)'	*person*	→	*pobl* 'people'

Since Welsh nouns are most often pluralized in category 1 and this is also the most common category to pluralize native English nouns in an English environment, it is not surprising to see that the vast majority of English-origin nouns are pluralized by adding a suffix. Category 3 shows a higher percentage for tokens than for types.

This is explained by the high token count of the noun *hogiau* 'boys' as it occurs 70 times. In the following, examples from all categories are listed:

Table 4: Examples for integrated English loanwords.

Category	Uniplex form	Multiplex form	Translation
Cat 1 (+suffix)	*ffrind*	*ffrindiau*	'friends' (from EME *frind(e)* 'friend(s)')
Cat 2 (+suffix + sound change)	*cwrs*	*cyrsiau*	'course' (from ME *course* 'course')
Cat 3 (~suffix)	*hogyn* *taten*	*hogiau* *tatws*	'boys' (from eng. *hogg* 'young animal') 'potatoes'
Cat 4 (~suffix + sound change)	*cerdyn* *cesyn*	*cardiau* *cases*	'card(s)' 'case(s)'
Cat 5 (collective)	*bricsen* *twlsyn* *pilsen*	*brics* *twls/ tools* *pills*	'brick(s)' 'tool(s)' 'pill(s)'
Cat 6 (collective + sound change)	*cyrensen* *rhecsyn*	*cyraints* *rhacs*	'currant(s)' 'rag(s)'
Cat 7 (sound change)	*ffordd* *plismon*	*ffyrdd* *plismyn*	'way(s)' (from OE *ford* 'ford') 'policeman/men'
Cat 8 (suppletive)	–	–	–

Perhaps the first thing that strikes the eye when reading the examples is the graphematic adaption of the loanwords. When Welsh number inflection is applied, there are only few cases in which the spelling stays identical with the original English noun (e.g. *problem-au* 'problem-s'). Consequently, a number of English loanwords are sometimes not recognized as non-Welsh at first glance. This may apply especially to nouns borrowed in the Old English and Middle English period. All of the 197 English loanwords are listed in GPC (cf. Deuchar et al. 2018 for the category 'listedness'). Therefore, the nouns presented in Table 4 and the other English nouns in the corpus applying Welsh number inflection are not spontaneous borrowings but are established lexemes of Modern Welsh.

Figures 6/7 and Table 4 provide a base for answering research question 2). There is a clear preference to add a suffix to the singular form of the noun (Cat. 1). This preference is more visible in English nouns, as 76% of types fall into category 1 compared to 63% Welsh noun types. Adding a suffix and a sound change (Cat. 2) is the appropriate strategy for 15% of types in both languages. Categories 3–6 each contain a very small number of types from both Welsh and English. Applying a vowel change (Cat. 7) is a strategy which is chosen less often in English loanwords than in Welsh nouns: 12% of Welsh nouns are pluralized this way, whereas only 2% of English loanwords apply this category. Category 8 is devoid of English loanwords from the Siarad corpus. In order to provide more details to this answer, the next two sub-sections zoom in on specific categories and the English loanwords they contain.

4.1 Collective nouns

Collective nouns (i.e. categories 5 and 6) have been discussed at length in various works. Stolz (2001) provides a list of English nouns which have been borrowed into the singulative/collective category and concludes that "massive language contact did not accelerate the expected disintegration of marked singulative-collective distinctions. On the contrary, the integration of English loan-words has even contributed to strengthening the system-internal role of singulative-collective distinctions" (Stolz 2001: 69).

According to Nurmio (2019), the rise of borrowed singulatives in the Early Modern Welsh period is part of a bigger picture, as the suffix *-yn/-en* fulfils several functions such as individuating and nominalizing, depending on their base. The suffix *-yn/-en* incorporates English nouns into Welsh and gives them a "more native form" (Nurmio 2019: 98).

Despite the former and potentially ongoing productivity of this category, borrowed collective and singulative nouns occur very infrequently in the Siarad corpus. Table 4 shows the complete list of English types in categories 5 and 6. Furthermore, the singulative counterparts are rare occurrences as well. The only singulative forms represented in the corpus are *bricsen* 'brick' (four tokens) and *twlsyn* 'tool' (one token). The low number of English-origin collective nouns in the Siarad corpus is not entirely unexpected, as collective nouns only make up a small percentage of Welsh-origin nouns as well (about 5% of tokens, see Figure 7).

Taking a quick glance at other corpora, we find a slightly higher number of English collective nouns in the Patagonia corpus (see Arbes 2022: 65). Despite Roberts & Gathercole's (2012: 82) claim that "collection[8] nouns make up the largest portion of the forms in Welsh that have multiple referents", the sample of the CEG corpus they consulted in fact contains 9% collective and 91% plural noun tokens.

---

[8] The term Collection/ Unit nouns employed in Roberts & Gathercole (2012) is congruent with the term collective/singulative used in this paper.

English borrowings, although they are rare, do occur in the collective category in CEG. The relevant examples are *betysen/ betys* 'beet/s', *bricsen/ brics* 'brick/s' and *cocsen/ cocos* 'cockle/s'.

A plethora of English nouns borrowed into the singulative/collective category is found in the considerably larger National Corpus of Contemporary Welsh (CorCenCC), e.g. *cetrisen/* cetris 'cartridge/s', *socsen/ socs* 'sock/s' and *bynsen/ byns* 'bun/s'.

The findings above show that in order to come across a reasonable number of collective and singulative nouns borrowed from English, a corpus with an extensive token count is required. Corpora encompassing a million tokens or fewer do not tend to include many examples of this much discussed phenomenon.

In the following sub-section, the category which contains most English loanwords is brought into focus.

4.2 Borrowed nouns in category 1

Plural nouns in category 1 (+suffix) occur sufficiently frequently for a quantitative analysis to be worthwhile. Do English borrowings show a preference for a specific suffix and is the frequency and distribution of suffixes the same for English and Welsh nouns? The answer is "yes" for the former and "no" for the latter question, as Figure 8 shows.

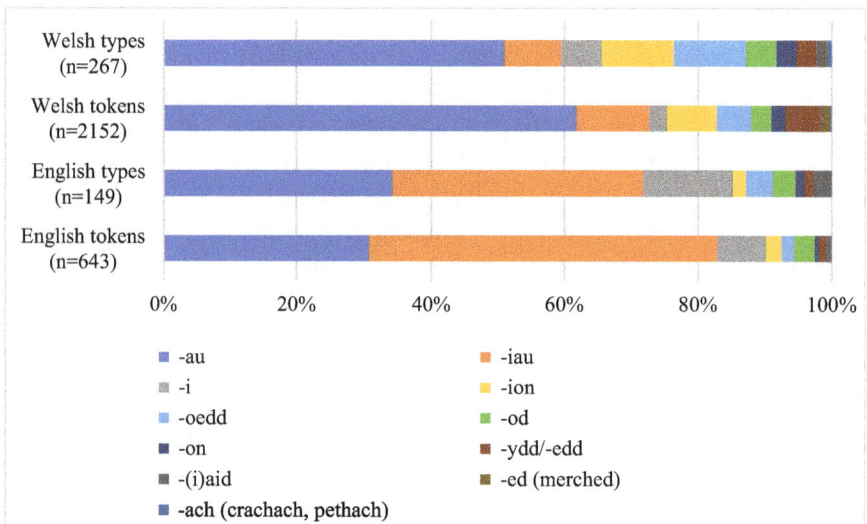

Figure 8: Share of suffixes in Category 1 (+suffix).

The suffix *-au* is added to Welsh nouns very often, in fact, about 51% of types and 62% of tokens in this category are inflected by *-au*. This preference is mirrored to some extent by the English borrowings, which apply *-au* to 34% of types and 30% of tokens. Common examples are e.g. *afal-au* 'apple-s' and *siop-au* 'shop-s'.

While the suffix *-iau* plays only a secondary role in Welsh native nouns, it is employed most frequently for English-origin nouns, as it is added to 37% of types and 52% of tokens. Welsh-origin examples include e.g. *llun-iau* 'picture-s' and *clust-iau* 'ear-s'; among English loanwords common examples are *ffilm-iau* 'film-s' and *grwp-iau* 'group-s'.

Similarly, although it occurs less frequently overall, the suffix *-i* is applied to English nouns more often than to Welsh nouns as well (e.g. *ticed-i* 'tickets', *sianel-i* 'channels'). The suffixes *-ion* and *-oedd* are quite frequent for Welsh nouns as each of them accounts for 11% of types, but in connection with English nouns they are rare. Cases found in the Siarad corpus are e.g. *actor-ion* 'actor-s' and *tim-oedd* 'team-s' (as an alternative to the more common form *tim-au*). In the following, English and Welsh nouns employing the suffixes with the highest frequency in this category, namely *-au* and *-iau*, are analyzed in more detail.

It has been established that the suffix *-iau* pluralizes mostly monosyllabic nouns (cf. Watkins & Wöbking 1992: 52). This is applicable to the Welsh nouns but even more clearly visible for the English borrowings in the corpus data, as 100% of English nouns pluralized by *-iau* are monosyllabic in their singular form (see also Figure 9).

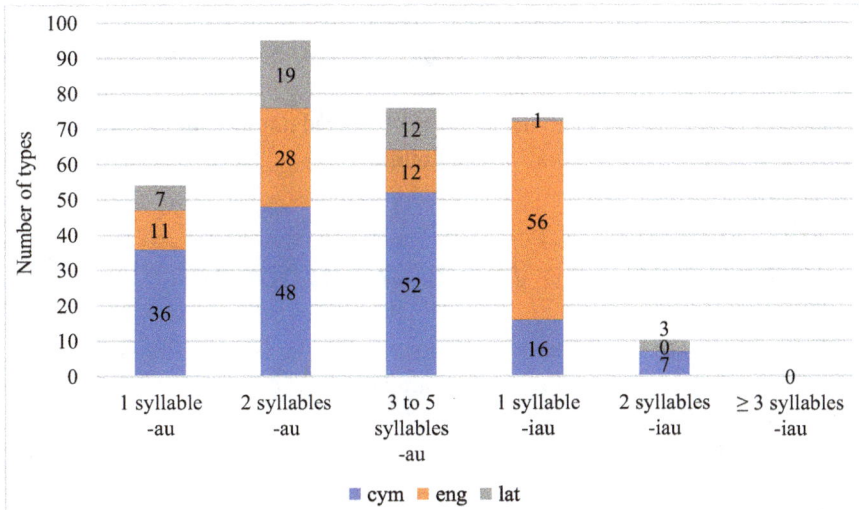

Figure 9: Syllable count of nouns pluralized by *-au* or *-iau* (n=308).

The opposite is true for the suffix *-au*, where nouns with more than one syllable are in the majority. However, it is not impossible to inflect a monosyllabic noun with the suffix *-au* (e.g. *timau* 'teams'). It is therefore more of a preference than a distinct rule to pluralize monosyllabic nouns with the suffix *-iau* rather than *-au*. While English nouns pluralized by *-iau* always consist of only one syllable in their singular form, not all monosyllabic nouns are pluralized by *-iau* (see first column in Figure 9).

The idea comes to mind that another variable besides language of origin and number of syllables may play a role. Awbery (1984) and Hannahs (2013) have established that in Welsh phonology the weight of a syllable (measured in moras) plays a role for word-formation and incorporating loans into the language. Examples with short vowels such as *mat-iau* [mat] 'mat-s' and *step-iau* [stɛp] 'step-s', compared to forms with a long vowel or diphthong such as *sioe-au* [ʃoːɨ̯/ ʃɔi̯] 'show-s' and *ceg-au* [keːg] 'mouth-s' led me to the hypothesis that monosyllabic nouns containing a short vowel are pluralized by *-iau*, whereas monosyllabic nouns containing a long vowel or diphthong are pluralized by *-au*. However, with this sample the null hypothesis could not be rejected. Statistical tests found no correlation between syllable weight and the distribution of *-au* and *-iau*, either in loanwords or in native Welsh nouns. A number of counter-examples to the ones mentioned above, e.g. *pwynt-iau* [puːɨ̯nt/ pui̯nt] 'point-s', *siap-iau* [ʃaːp] 'shape-s' and *swm-au* [sʊm] 'sums' illustrate this result.

Generally, English nouns differ phonologically from Welsh nouns in various ways (cf. Awbery 1984). In the sub-sample of nouns pluralized by *-au* and *-iau* the Welsh nouns contain more syllables on average than the English nouns. English monosyllabic nouns differ from Welsh nouns in that they contain short vowels more often than long vowels or diphthongs. For the categories of number inflection this may mean that they come to incorporate borrowed nouns which do not exactly match the patterns of the familiar native nouns.

5 Plural suffix –*s* on Welsh-origin nouns?

In order to answer research question 3 "How frequently is the suffix *-s* borrowed and applied to Welsh-origin nouns in the Siarad corpus?" a list of all relevant nouns is provided in Table 5.

The table shows a mixture of attested forms and supposed spontaneous creations. *Gogs* 'northerners' and a differently spelled form of *hearns* 'irons' are mentioned in GPC. The GPC entries show a great variation of plural options in some cases, some of which were also used by speakers in the Siarad corpus (e.g. *bleiddi-*

Table 5: Welsh nouns pluralized by –s.

Pluralized form with -s attested in the Siarad corpus	Version attested in GPC	Translation	Tokens
gogs	gogs, gogleddwyr	'people who live in North Wales'	17
bais	beiau, beion, beioedd	'blames'	1
blaidds	bleiddiaid, bleiddiau, bleiddau, bleiddiawr, bleiddawr	'wolves'	1
cadw-mi-geis	–	'piggy-banks'	1
cefnders	cefndyr (gynt cefndyrw), cefndryd, cefnderwyr, cefnderoedd, cefnderwydd, cefnderwedd, cefnderon	'(male) cousins'	1
chwertins	–	'laughs'	1
crancs	crancod, crangod, crainc	'crabs'	1
gr_s	–	'consonant clusters <gr>'	1
haearns	heyrn, haearnau, heyrns	'irons'	1
lan-môrs	glannau, glennydd	'sea-shores'	1
r_s	–	'letters <r>'	1
taids	teid(i)au, teidoedd	'grandfathers'	1

aid 'wolves') and *cefndryd* 'cousins'). Even though more modern dictionaries may list fewer variants, this shows that historically, there has been no clear consensus[9] about the plural form. A possible explanation for the forms pluralized by -s is that the Welsh plural form was unknown or had been forgotten and the suffix -s came to mind as an alternative.

It has been claimed that nouns may be pluralized by -s especially in informal settings (cf. P. W. Thomas 1996:175) and in some cases, the corpus data obviously shows that the speakers perceived the conversations as informal. In example (12) the participants laugh and make helicopter sounds and then wonder how the transcriber will put their sounds into writing.

(12)   Davies-09, (201) LLE: [laughing, helicopter sounds],
    be   *mae*   *wneud* *efo*   *chwerthins* *yma* *ta*
    what be.3SG.PRES do.VN   with laughter:PL there then
    'what's he doing with these laughs then?'

---

9 For a more detailed analysis of nouns with multiple plural forms see Stolz (2008).

Although not all interviews in the Siarad corpus are accompanied by laughter and silliness, examples such as this one show that the creators of the Siarad corpus have provided a space for informal conversation where it became possible to find forms which would not have been observable in written corpora or in more formal settings.

In comparison, the CEG corpus, which consists of over one million words of written prose texts, only contains a handful of Welsh nouns pluralized by -s such as *pregethwrs* 'preachers' and *dengwrs* 'ten men'.

While it is shown that the suffix -s may pluralize Welsh nouns on occasion, especially in spoken and informal language, it is important to note that compared to nouns with Welsh number inflection, the forms employing -s are still extremely rare as they account for less than 3% of types and less than 1% of tokens.

6 Conclusions

In the following, the answers to the research questions proposed in Section 1) are summarized and possibilities for further research are discussed. First, English nouns pluralized by the suffix -s were discussed. While these single-word code-switches account for only 18% of multiplex tokens, their frequency exceeds that of English loanwords with Welsh number inflection. Most of them do not adhere to the rules of initial mutation, but the corpus also shows that it is possible for otherwise morphologically unintegrated nouns to undergo a soft mutation.

English loanwords show some differences in their distribution across categories for number inflection when compared to Welsh-origin nouns. As expected, English nouns occur more often in category 1 (+suffix) and significantly less often in category 7 (vowel change). The only category devoid of English borrowings is the suppletive category (8). A further view into category 1 has shown that differences are visible on the level of individual suffixes as well. English nouns, especially monosyllabic ones, show a preference for the suffix -*iau*, while Welsh nouns in this category are most often pluralized by -*au*. The suffix -*au* comes second place for English loanwords and is followed by -*i*. The suffixes -*ion* and -*oedd*, which are each responsible for 11% of Welsh pluralized types in category 1 only incorporate a small number of English loanwords.

Regarding Welsh nouns and potential borrowing of the English plural suffix -s, the Siarad corpus shows that these forms exist, especially in informal settings. However, they still constitute an exception and thus the Welsh suffixes are by no means endangered.

The Siarad corpus is the first of its kind and has led to some eye-opening results. Topics such as codeswitching, which were mostly speculated about based on anecdotal evidence, have since been researched quantitatively and qualitatively on a much more solid basis. It is important to keep in mind that the data from the Siarad corpus is a snapshot of spoken Welsh between 2005 and 2008. Future studies may take newer corpora into consideration and assess the possible changes in the use of Welsh between decades. Since this paper focusses on number inflection, other word classes were only mentioned in passing, which is not to say they are unimportant. A still unanswered question is how the frequency of plural nouns compares to that of other parts of speech regarding codeswitching. The relation between register and codeswitching is another topic deserving of more attention.

Number inflection of Welsh and borrowed nouns is a topic I will continue researching. Issues such as multiple plural forms, productivity of individual suffixes and collective/ singulative nouns, which played a minor role in this paper, will be discussed further in upcoming projects.

Abbreviations

1,2,3	first/ second/third person
CorCenCC	Corpws Cenedleithol Cymru (National Corpus of Welsh)
cym	Cymraeg/ Welsh
DET	determiner
EME	Early Modern English
eng	English
GPC	Geiriadur Prifysgol Cymru (A Dictionary of the Welsh Language)
IPFV	imperfective
L	lenition
lat	Latin
ME	Middle English
MLF	Matrix Language Frame
OE	Old English
PL	plural
PRES	present
PST	past
PTL	particle
SG	singular
VN	verbal noun

Primary Sources

Deuchar, Margaret. 2011. *Bangor Patagonia Corpus*. http://bangortalk.org.uk (14 February, 2018).
Deuchar, Margaret & Peredur Davies. 2014. *Bangor Siarad Corpus*. http://bangortalk.org.uk (14 February, 2018).
Ellis, N. C., C. O'Dochartaigh, W. Hicks, M. Morgan & N. Laporte. 2001. *Cronfa Electroneg o Gymraeg (CEG): A 1 million word lexical database and frequency count for Welsh*. www.bangor.ac.uk/canolfanbedwyr/ceg.php.en (6 November, 2020).
Knight, Dawn, Steve Morris, Tess Fitzpatrick, Paul Rayson, Irena Spasić, Enlli M. Thomas, Alex Lovell, Jonathan Morris, Jeremy Evas, Mark Stonelake, Laura Arman, Josh Davies, Ignatius Ezeani, Steven Neale, Jennifer Needs, Scott Piao, Mair Rees, Gareth Watkins, Lowri Williams, Vignesh Muralidaran, Bethan Tovey-Walsh, Laurence Anthony, Thomas M. Cobb, Margaret Deuchar, Kevin Donnelly, Michael McCarthy & Kevin Scannell. 2020. *CorCenCC: Corpws Cenedlaethol Cymraeg Cyfoes – the National Corpus of Contemporary Welsh*: Cardiff University.
Thomas, R. J., Gareth A. Bevan, Patrick J. Donovan & Andrew Hawke (eds.). 1967–2002. *Geiriadur Prifysgol Cymru/A Dictionary of the Welsh Language*. Cardiff: University of Wales Press. https://geiriadur.ac.uk/gpc/gpc.html (06.05.2021)

References

Arbes, Deborah. 2022. Language contact and number inflection in Patagonian Welsh. In Nataliya Levkovych (ed.), *Susceptibility vs. Resistance: Case studies on different structural categories in language-contact situations*, 51–89. Berlin & Boston: De Gruyter Mouton.
Awbery, Gwenllian M. 1984. Phonotactic Constraints in Welsh. In Martin J. Ball & Glyn E. Jones (eds.), *Welsh phonology: Selected readings*, 65–104. Cardiff: University of Wales Press.
Binks, Hanna. 2017. *Investigating the bilingual 'catch-up' in Welsh-English blingual teenagers*. Thesis (Ph.D.). Bangor.
Carter, Diana, Margaret Deuchar, Peredur Davies, María del Carmen Parafita Couto. 2011. A Systematic Comparison of Factors affecting the Choice of Matrix Language in Three Bilingual Communities. *Journal of Language Contact*(4), 153–183.
Deuchar, Margaret. 2006. Welsh-English code-switching and the Matrix Language Frame model. *Lingua* 116(11). 1986–2011.
Deuchar, Margaret & Peredur Davies. 2009. Code switching and the future of the Welsh language. *International Journal of the Sociology of Language* 2009(195). 15–31.
Deuchar, Margaret, Kevin Donnelly & Caroline Piercy. 2016. 'Mae pobl monolingual yn minority': Factors Favouring the Production of Code Switching by Welsh-English Bilingual Speakers. In Mercedes Durham & Jonathan Morris (eds.), *Sociolinguistics in Wales*, 209–239. London: Palgrave Macmillan.
Deuchar, Margaret, Kevin Donnelly & Peredur Webb-Davies. 2018. *Building and Using the Siarad Corpus*: *Bilingual Conversations in Welsh and English* (Studies in Corpus Linguistics 81). Amsterdam/Philadelphia: John Benjamins Publishing Company.
Gardani, Francesco. 2015. Plural across inflection and derivation, fusion and agglutination. In Francesco Gardani, Peter Arkadiev & Nino Amiridze (eds.), *Borrowed Morphology*, 71–97. (Language Contact and Bilingualism 8). Boston: De Gruyter Mouton.

Hannahs, Stephen J. 2013. *The phonology of Welsh* (The phonology of the world's languages). Oxford: Oxford University Press.

Haspelmath, Martin. 2009. Lexical borrowing: concepts and issues. In Martin Haspelmath & Uri Tadmor (eds.), *Loanwords in the World's Languages*, 35–54. Berlin, New York: Walter de Gruyter.

Haspelmath, Martin & Andres Karjus. 2017. Explaining asymmetries in number marking: Singulatives, pluratives, and usage frequency. *Linguistics* 55(6). 1213–1235.

Jones, Mari C. 1998. *Language obsolescence and revitalization: Linguistic change in two sociolinguistically contrasting Welsh communities* (Oxford studies in language contact). Oxford: Clarendon Press.

King, Gareth. 2003. *Modern Welsh: A comprehensive grammar.* (Routledge Grammars). London: Routledge.

Lloyd-Jones, J. 1910. Some Latin Loan-Words in Welsh. *Zeitschrift für celtische Philologie* 7(1). 462–474.

Matras, Yaron. 2020. *Language contact.* (Cambridge textbooks in linguistics). Cambridge, United Kingdom, New York, NY, USA: Cambridge University Press.

Myers-Scotton, Carol. 1998. A way to dusty death: the Matrix Language turnover hypothesis. In Lenore A. Grenoble & Lindsay J. Whaley (eds.), *Endangered Languages*, 289–316. Cambridge: Cambridge University Press.

Myers-Scotton, Carol. 2002. *Contact Linguistics: Bilingual Encounters and Grammatical Outcomes.* (Oxford linguistics). Oxford: Oxford University Press.

Nurmio, Silva. 2017. Collective Nouns in Welsh: A Noun Category or a Plural Allomorph? *Transactions of the Philological Society* 115(1). 58–78.

Nurmio, Silva. 2019. Grammatical Numer in Welsh: Diachrony and Typology 117(S1).

Parina, Elena. 2010. Loanwords in Welsh: Frequency analysis on the basis of Cronfa Electroneg o Gymraeg. In Dunja Brozovic Roncevic, Maxim Fomin, Ranko Matasovi (ed.), *Celts and Slavs in Central and Southeastern Europe. Studia Celto-Slavica III: Proceedings of the IIIrd International Colloquium of the Societas Celto-Slavica*, 183–194. Zagreb: Institut za hrvatski jezik i jezikoslovlje.

Parry-Williams, Thomas H. 1923. *The English Element in Welsh: A study of English loan-words in Welsh.* London: The honourable society of Cymmrodorion.

Roberts, Anna E. 1988. Age-related variation in the Welsh dialect of Pwllheli. In Martin J. Ball (ed.), *The Use of Welsh: A contribution to sociolinguistics*, 104–122. (Multilingual matters 36). Clevedon, England: Multilingual Matters.

Roberts, Seren & Virginia Gathercole. 2012. Talking of Objects: How Different are Welsh and English Nouns? *Journal of Celtic Linguistics* 14(1). 67–85.

Stammers, Jonathan & Margaret Deuchar. 2012. Testing the nonce borrowing hypothesis: Counter-evidence from English-origin verbs in Welsh. *Bilingualism: Language and Cognition* 15(3). 630–643.

Stolz, Thomas. 2001. Singulative-Collective: Natural Morphology and stable classes in Welsh number inflexion on nouns. *STUF – Language Typology and Universals* 54(1). 52–76.

Stolz, Thomas. 2008. Kymrische Ausnahmen oder walisische Regeln? Was die substantivische Pluralvariation uns lehrt. In Cornelia Stroh (ed.), *Morphologische Irregularität: Neue Ansätze, Sichtweisen und Daten*, 111–150. (Diversitas linguarum 19). Bochum: Brockmeyer.

Thomas, Alan R. 1987. A spoken standard for Welsh: description and pedagogy. *International Journal of the Sociology of Language* 1987(66). 99–113.

Thomas, Enlli M. Nia Willimas, Llinos Angharad Jones & Hanna B. Susi Davies. 2014. Acquiring complex structures under minority language conditions: Bilingual acquisition of plural morphology in Welsh. *Bilingualism: Language and Cognition* 17(03). 478–494.

Thomas, Peter W. 1996. *Gramadeg y Gymraeg.* Caerdydd: Gwasg Prifysgol Cymru.

Thomason, Sarah G. 2015. When is the diffusion of inflectional morphology not dispreferred? In Francesco Gardani, Peter Arkadiev & Nino Amiridze (eds.), *Borrowed Morphology*, 27–46. (Language Contact and Bilingualism 8). Boston: De Gruyter Mouton.

Watkins, Thomas A. & Konstantin Wöbking. 1992. *Kurze Beschreibung des Kymrischen.* (Innsbrucker Beiträge zur Sprachwissenschaft 71). Innsbruck: Institut für Sprachwissenschaft.

Alexandre Arkhipov

3 Two morphologies, two stress systems, shaken, not stirred

Number marking on Russian borrowings in Kamas

Abstract: The paper investigates strategies of adaptation of Russian loans in Kamas, an extinct Samoyedic (< Uralic) language of Southern Siberia. The study is based on the INEL Kamas Corpus which includes transcripts of all the audio data recorded from the last speaker between 1963 and 1970. Three main strategies of number marking on borrowed nouns are observed, two of them involving a Kamas plural marker attached to a Russian singular or plural form, and the third one directly employing the Russian plural. In a few more intricate cases, various adaptations take place, sometimes resulting in forms not expected in any of the interacting languages by itself. One minor strategy involves a stress shift in Russian singular form, another one uses an oblique stem coinciding with Russian genitive singular.

Keywords: borrowings, nouns, number, stress, vowel lengthening, tonal accent, Kamas, Russian, Turkic

1 Language and data

1.1 Kamas language

Kamas belongs to the Samoyedic branch of the Uralic language family. All the main sources document the Forest Kamas varieties spoken in the settlement of Abalakovo, in the present Krasnoyarsk Krai in Southern Siberia, on the northern slopes of the East Sayan Mountains. The language became extinct by the late XXth century

Acknowledgements: This publication has been produced in the context of the joint research funding of the German Federal Government and Federal States in the Academies' Programme, with funding from the Federal Ministry of Education and Research and the Free and Hanseatic City of Hamburg. The Academies' Programme is coordinated by the Union of the German Academies of Sciences and Humanities.

I am grateful to Valentin Gusev, Gerson Klumpp and Sergey Knyazev for reading the first draft of this paper and for extensive and fruitful discussions of the data. Many thanks to Isabel Compes, Katharina Gayler and Elsadig Omda Ibrahim Elnur, as well as to Deborah Arbes, the editor of this volume, for very thorough reviews which helped to make this paper much clearer and more consistent. All remaining shortcomings and errors are mine.

https://doi.org/10.1515/9783110986600-003

with the death of its last known speaker, Klavdiya Plotnikova, in 1989. It was documented by several researchers during XVIIIth–XXth centuries, however, substantial Kamas texts were only recorded in the XXth century.

Kamas is morphologically agglutinating with some elements of fusion, predominantly suffixing. Nouns have two declension types, absolute and possessive declension. In verbal morphology subjective and objective conjugation are distinguished, the latter as an incomplete paradigm. Kamas is rich in derivational suffixes, especially in the verbal domain. Its phonology and lexicon, as well as such grammatical features as converb constructions and aspectual auxiliaries, show impact of contact with Turkic languages. Some of the Kamas speakers eventually shifted to a local Turkic variety (Kacha dialect of Khakas), while the others finally shifted to Russian in the first half of the XXth century.

Kai Donner, an eminent Finnish linguist, worked with Kamas speakers in 1912 and 1914. The materials he collected have been posthumously edited and published by Aulis J. Joki (1944). According to Donner, the language proficiency in the community was already declining and the language was not as rich and elaborate as it used to be; already at that time the youngest more or less fluent speaker was over 45 (Joki 1944: XLI). A few years after Donner's visits, the language was thought to be extinct. Nonetheless, in 1963 a surviving Kamas speaker, Klavdiya Plotnikova, born in 1895, was discovered during a toponymic fieldtrip by Aleksandr Matveyev and his students in Abalakovo. By that time, she had not spoken her language for some 20 years. However, she still remembered the language to some extent, and reactivated its use in subsequent years while working with linguists. She was recorded on tape in 1963–1970, mainly by the Estonian scholar Ago Künnap. Two shorter recordings were made around 1964 from another speaker, Aleksandra Semyonova, in Krasnoyarsk; she died soon afterwards.

Both Plotnikova and Semyonova had most probably not fully acquired Kamas in their childhood and were recorded after many years of not using the language. As Klumpp (2022b: 27) points out, their gaps in Kamas performance "are due to attrition of the *forgetter* type, i.e. forgotten material and structure, and to the *semi-speaker* type, i.e. never acquired material and structure." Both are also reported by Matveyev (1965: 34, 37) to speak Kacha Khakas to some extent. While for Plotnikova this is only mentioned, some Khakas words and phrases are recorded in Semyonova's tapes. The post-shift Kamas is a product of intensive erosion, with very restricted grammar and lexicon, inconsistent use of grammatical markers and many speech disfluencies, showing Russian influence on all language levels. Nevertheless, the amount of speech recorded from Plotnikova is significant (ca. 14 hours including Russian) and provides valuable insights into the otherwise scarcely documented language.

1.2 Data used in this study

The present study is based on the INEL Kamas Corpus (Gusev et al. 2019),[1] which includes transcripts, interlinear glosses and further annotations for the entire body of the available sound recordings of the post-shift Kamas speech. For details on the corpus composition and annotations, see (Arkhipov et al. 2020); for the general presentation of the INEL project, see (Arkhipov & Dabritz 2018).

The total corpus size is 63,824 tokens (words), including 2,500 words in the pre-shift part (Donner's collection) and 61,324 words in the post-shift part. Of these, ca. 14,700 words belong to utterances of the interviewers and to Plotnikova's Russian speech; thus the Kamas proper part of the post-shift corpus amounts to ca. 46,600 tokens.

The examples studied here are 258 tokens[2] of Russian loan nominals bearing one (or more) of the available plural markers, along with other grammatical forms of the same lexemes. 5 of the 258 tokens are morphologically adjectives in the donor language, the rest are nouns. The same and other nouns occurring in utterances spoken entirely in Russian are not counted here, but sometimes referred to for comparison.

It should be kept in mind that the transcriptions, glosses and translations of the examples are not identical to those in the currently published version of the corpus (1.0). On the one hand, the examples in focus were double-checked and transcriptions corrected in several cases, especially with respect to the vowels and such consonantal features as voicing and palatalization. Glosses were sometimes corrected as well to achieve more consistency in the reflection of the observed plural marking strategies. These corrections will be included in an amended version of the corpus. On the other hand, the transcriptions given here were enhanced to reflect stress and vowel lengthening in Russian loans, otherwise unmarked in the corpus. Translations of sentences were adapted for clarity.

---

1 The corpus can be downloaded at http://hdl.handle.net/11022/0000-0007-DA6E-9. An online search interface is available at https://inel.corpora.uni-hamburg.de/KamasCorpus/search. See also the project page at https://inel.corpora.uni-hamburg.de/.

2 While I do not list exact references (text names and sentence numbers) for the examined word forms, most of them can be retrieved and played back online by searching for grammatical features "n,pl" (i.e. "noun, plural") in the "Grammar" field, and "(RUS:core | RUS:cult)" in the "Borrowing" field at https://inel.corpora.uni-hamburg.de/KamasCorpus/search. A corrected version of the corpus is to be published by the end of December 2023.

1.3 Borrowings in Kamas

Two main sources of borrowings in Kamas are South Siberian Turkic languages, notably Kacha Khakas, and Russian. Extensive contacts with the neighboring Turkic languages are responsible not only for lexical borrowings but also for structural ones, including converb constructions and aspectual auxiliaries. These are well documented in pre-shift Kamas.

Russian borrowings in Donner's texts are, in contrast, very limited. Many of them are conjunctions: *i* 'and', *da* 'and', *a* 'and, but', *al'i* 'or'and *kak* 'as'. In the post-shift corpus, conjunctions are the most frequent of Russian borrowings as well.

Table 1 displays the counts of Russian and Turkic borrowings in the post-shift part of the INEL Kamas Corpus. Although the annotation of borrowings in the corpus is not exhaustive and might not be fully precise, it is representative enough to demonstrate the main tendencies. Most borrowings in the 'lexical' class are nouns, with just a few verbs, adjectives and adverbs. The other three classes regroup function words and morphemes, including grammatical devices such as conjunctions, discourse particles and modal words.

Table 1: Turkic and Russian borrowings in the post-shift part of the Kamas corpus.[3]

Class	Turkic			Russian		
	Tokens	%	Types	Tokens	%	Types
lexical	2384	61.3%	37	1382	25,5%	476
grammatical	262	6.7%	1	3076	56,8%	27
discourse	1218	31.3%	2	213	3,9%	10
modal	24	0.6%	2	750	13,8%	29
Total	3888	100%	41	5421	100%	542

Turkic lexical loans are deeper entrenched in the core lexicon. Although being much less numerous than Russian ones (resp. 37 vs. 476 lexemes), they are far more frequent in the corpus (2384 vs. 1382 tokens). This includes, for instance, the verbs *d'ăbaktər-* 'speak' and *togonər-* 'work', adjectives *jakšə* 'good' and *baška* 'other'. Among the 18 most frequent borrowed nouns listed in Table 2, Turkic loans occupy 13 lines including #1 to #9.

---

3 As of the time of writing, the current published version of the corpus is 1.0. Counts in Tables 1.1–1.2 are based on an updated working version of the corpus. An update is scheduled to be published by the end of 2023.

Of the remaining non-lexical borrowings, it is the polyfunctional particle *bar* 'all; (emphatic particle)' which is alone responsible for 1151 of 1504 occurrences.

Table 2: Most frequently borrowed nouns in the post-shift part of the Kamas corpus.

Rank	Word	Gloss	Count	Origin	Rank	Word	Gloss	Count	Origin
1	*tura*	house	223	Turkic	10	*volk*	wolf	54	Russian
2	*il*	people	220	Turkic	11	*p'eːš*	stove	52	Russian
3	*ipek*	bread	148	Turkic	12	*baltu*	axe	43	Turkic
4	*aba*	father	118	Turkic	13	*kujnek*	shirt	41	Turkic
5	*tüžöj*	cow	115	Turkic	14	*süt*	milk	38	Turkic
6	*kudaj*	god	94	Turkic	15	*mašina*	car	34	Russian
7	*akt'a*	money	93	Turkic	16	*s'estra*	sister	33	Russian
8	*ara*	vodka	91	Turkic	17	*t'erman*	hand mill	31	Turkic
9	*koŋ*	chief	57	Turkic	18	*kuris*	chicken	30	Russian

The Russian input is most visible in the grammatical devices: three out of a dozen loan conjunctions, *i* 'and', *a* 'and, but' and *da* 'and', account for almost 2,400 tokens of this class. Let us also mention the suffix *-n'ibud'* forming indefinite pronouns, e.g. *ĭmbi-n'ibud'* 'something' < *ĭmbi* 'what' (cf. Rus. *čto-n'ibud'* 'something' < *čto* 'what'), *gijən-n'ibud'* 'somewhere' < *gijən* 'where'.[4] A wide range of modal words is also prominently represented, such as *nada* 'one should', *možet* 'maybe', *axota* 'one wants', as well as focus particles like *tože* 'also', *iš:o* 'still; yet; more', *tol'ko* 'only'.

A final remark is due: what was discussed in this section only concerns borrowed material, not the structural borrowings (grammatical patterns), which are also well represented in the corpus.

1.4 Goal and structure of this study

The main goal of this study is to examine different strategies of nominal plural formation found in Russian loans in Kamas.

P. Roseano (2014) identifies three kinds of strategies that synthetic languages can use to form the plural of borrowed nouns, which he calls 'oikomorphological', 'xenomorphological' and 'allomorphological'. He finds that in Friulian, all the three kinds of strategies coexist and can occur on one and the same noun.

---

4 The symbol '<' is used in the sense of 'originating from', both for language-internal evolution or derivation and for borrowing processes.

Oikomorphological strategies correspond to the highest level of morphological integration: borrowings behave just as native words, using the markers of the target language according to the same rules as native words. Xenomorphological strategies, in contrast, correspond to the lowest level of morphological integration: borrowings retain their plural as in the donor language. Finally, allomorphological strategies are intermediate, including all cases where the plural formation deviates both from the donor language and from the target language.

As we will see, the two main parameters relevant for Russian loans in Kamas are (i) the grammatical form of the Russian noun (referred to as 'stem' hereafter) used to construct the resulting Kamas plural form, and (ii) the presence/absence and choice of the Kamas plural marker attached to that stem.

While the three most common strategies can be straightforwardly characterized according to these parameters, there is a handful of special cases where it is not that easy to identify the Russian stem involved. In order to solve the puzzle, we will need to dive into questions of word prosody and phonemic systems of the two languages.

We will proceed with necessary background information on the structure of Kamas (§2) and of Russian (§3) with respect to phonemic inventory, nominal morphology and stress, since these will all be relevant for the issues in focus. Some basic data on Russian dialects and specifically on the variety spoken by Plotnikova is also in order (§4). The main section (§5) reviews the strategies of plural formation found in Russian loans in Plotnikova's recordings, from the most common patterns to the most rare and problematic ones. The last section (§6) summarizes the main findings.

1.5 Disclaimers

The analysis presented in this paper is limited in several respects.

The Kamas spoken by Klavdiya Plotnikova cannot be identified with the Kamas of earlier times and must be regarded as a standalone phenomenon, whose properties can only be extrapolated to earlier, pre-shift stages of Kamas with a greater or lesser degree of plausibility.

The first reason is evidently the drastic language attrition and the impact of the recently dominant language, i.e. Russian, which make the difference between pre-shift and post-shift Kamas. The second reason is the relatively low generalization possibilities. On the one hand, it must be recognized that Plotnikova's Kamas is inherently variable, in that one and the same lexical item, grammatical form or construction can be realized differently in different parts of the corpus and even in adjacent utterances. On the other hand, the recording sessions contain a great deal of speech disfluencies such as false starts and self-repairs, so that slips of the tongue or otherwise unintended production cannot be ruled out as a potential explanation for unexpected forms, although I try to exclude the cases most susceptible of such perturbations from the analysis.

The variety of Russian spoken by Klavdiya Plotnikova, in turn, is not identical with the Modern Literary Russian (*СРЛЯ*, or Standard Russian; Std. Rus. or just Rus. for short). It is undoubtedly one of the Siberian regional varieties, however, a detailed dialectological characterization is a difficult task. Siberian varieties themselves represent complex products of interaction of many population groups with different dialectal background (see §3.2). As for Plotnikova's Russian, not only the amount of recorded Russian speech is fewer than that of her Kamas speech, but also her Russian is as well subject to variation. In particular, stress placement is variable in some lexemes: cf. *vólk'i ~ volk'í* 'wolves',[5] Std. Rus. *vólk'i; dóč'ka ~ doč'ká* 'daughter',[6] Std. Rus. *dóč'ka*. Such internal variation means that concurrent underlying Russian forms might stand behind particular shapes of a borrowed noun in Kamas.

That being said, I will still attempt to address the peculiarities of a limited fragment of grammar in Plotnikova's Kamas, to the extent that regular patterns of interaction between a (reconstructed) Kamas system and a (reconstructed) dialectal Russian system can be observed in the data.

2 Preliminaries on Kamas

2.1 Kamas phonology

Most existing descriptions of the Kamas phonology target its pre-shift state as documented by M. A. Castrén (1847) and Kai Donner in 1912 and 1914 (Joki 1944). Some unresolved problems remain, e.g. concerning the status of vowel length and the non-high front vowels. The following brief account is based on Klumpp (2022a: 821–823), except remarks on Plotnikova's data; some simplifications were made, and transcription symbols adapted to match the (non-IPA) transcription used in the INEL Kamas Corpus.

---

5 PKZ_19700821_09340-1bz.PKZ.044, 074, 088. References to examples from the INEL Kamas Corpus include: main speaker code(s) (here 'PKZ'); date of recording ('19700821'); text short title/code ('09340-1bz'); genre (e.g. 'flk' for folklore, 'nar' for narratives, here unspecified); in dialogues, specific speaker code ('PKZ'); sentence number ('044').
6 PKZ_19700819_09343-1a.PKZ.020, 002.

The Kamas consonant system includes a contrast in palatalization, marked in Table 3 with an apostrophe. A notable difference from Russian (cf. Table 8 below) is the presence of glottals and of the velar nasal in Kamas. The exact phonetic nature of the fortis/lenis (or voiceless/voiced) contrast (e.g. /p/ vs. /b/) is not clear, as is the phonemic status in some palatalized/non-palatalized and fortis/lenis pairs (given in square brackets in Table 3). Several phonemes given here in brackets only occur in loans.

Table 3: Pre-shift Kamas consonant system (after Klumpp 2022a: Table 39.3, 822).

	labial		dental/alveolar		postalveolar		palatal		velar		glottal
stops	p p'	b b'	t [t'][7]	[d d'][8]			t'[8]	[d'][8]	k k'	g [g']	ʔ
fricatives	(f)		s s'	[z z']	š š'	[ž]					h h'
affricates			(c)		(č')						
nasals	m m'		n n'							ŋ	
laterals			l [l']								
trills			r [r']								
approx.								j			

Full vowels, as shown in Table 4, form a quadrilateral system similar to those of the neighboring Turkic languages such as Khakas. The high back unrounded /ɨ/ only occurs in Turkic and Russian loans. The front rounded vowels /ü ö/ are pronounced as "rather central" (IPA [ʉ ɵ]) (Klumpp 2022a: 821). The distinctions between /u o/ and /ü ö/ in Plotnikova's pronunciation are not entirely consistent and need further investigation; when phonetic transcription is needed, I will provisionally transcribe both pairs as [u o] in most cases.

Some lexical items contain vowels with extremely variable realizations; these are taken to represent 'reduced' vowel phonemes. E.g., some forms of the root /băt-/ 'cut' are recorded as [bɔð-] ~ [boð-] ~ [buð-]. In first syllables, only the front–back distinction is kept in reduced vowels (/ĭ/ vs. /ă/). In non-first syllables, this distinction is also neutralized in /ə/, which can be realized as [ə] or assimilate to a preceding vowel, e.g. /bübə/ [bʉbʉ] 'his/her water [ACC]' (Klumpp 2022a: 821).

In Plotnikova's post-shift data, /ə/ is regularly rounded after a rounded vowel in the preceding syllable, except if an earlier long/stressed vowel in the root is unrounded, cf.: *kömə* [komu] 'red', *šü-gən* [šuɣun] 'fire-LOC', *tüžöj-əʔí* [tužoːjuʔí] 'cow-PL'; but *xaːr'uz-əʔí* [xaːr'uzəʔjí] 'grayling-PL' (< Rus. *xár'ius*). This rounding

---

7 Note that palatal stops (IPA /c ɟ/) are represented in the INEL Kamas Corpus with /t' d'/, same as palatalized coronal stops (IPA /tʲ dʲ/) in Russian loans. Hushing sibilants (IPA /ʃ ʒ/) are written as /š ž/, and affricates (IPA /ts tʃ/) as /c č/.

assimilation can optionally be carried on to a following /ə/, e.g. *plod-əʔjə* [plo:duʔjú] 'raft-PL' (< Rus. *plot*).

In §5, I will reflect this schwa rounding in transcription of epenthetic /ə/ in suffixes (see §2.2) for disambiguation – although Russian stem-final /a/ or /i/ can be realized as [ə] in radical reduction, they do not become [u] after rounded vowels.

Table 4: Pre-shift Kamas vowel system (after Klumpp 2022a: Table 39.2, 821).

Vowels		front		back	
		unrounded	rounded	unrounded	rounded
full	high	i	ü	(ɨ)[8]	u
	mid	e	ö		o
	low			a	
reduced	1st syllable		ĭ		ă
	non-1st sylllable			ə	

Vowel length is phonemic when resulting from contractions, e.g. /kan-/ 'go' vs. /ka:n/ 'Khan' (< Kacha Khakas *qayan* 'Khan'), *küla:mbi* '(s)he died' (< *kü-le kambi* '(lit.) dying went away'). It is non-phonemic in a few other cases where lengthening is automatic. Non-high vowels[9] are automatically lengthened: firstly, in open syllables resulting from suffixation (/oro/ [oro] 'pit' > /oro-nə/ [oro:nə] 'pit-LAT'; /šürget/ [šʉrget] 'elbow' > /šürged-ən/ [šʉrge:dən] 'elbow-GEN'); and, secondly, before /r/ (/par-/ [pa:r-] 'return'). The originally stressed vowels in Russian loans are also typically lengthened, e.g. /koška/ [ko:ška] 'cat' (< Rus. *kóška*) (Klumpp 2022a: 821).

There is no lexical tone in Kamas.

2.2 Kamas nominal morphology

Kamas nouns inflect for case, number and possession. The so-called absolute declension (illustrated in Table 5) is a sub-paradigm with case and number inflection only. The possessive declension involves possession, case and number (Table 6). Seven cases include zero-marked nominative, accusative, genitive, lative, locative, ablative and instrumental. Examples in this section are adapted from Klumpp (2022a): 824–827.

---

8 The phoneme /ɨ/ indicated here in brackets only appears in loans.
9 Klumpp (2022: 821) speaks about 'open vowels' here, but his examples actually include /e o/ which he labels in the vowel chart as 'open/close mid'. I will use the term 'non-high vowels' to refer to the same class, which includes 'mid' and 'low' in Table 4.

Singular number is zero-marked. Dual only appears in possessive declension (and only for possessors), on pronouns and in verbal agreement (not treated here). Plural has two markers, the older *-jaʔ ~ -iʔ* (in Plotnikova's recordings, *-ʔi ~ -ʔjə*) and the newer *-zaŋ/-zeŋ* (*-saŋ/-seŋ* after obstruents); the harmonic alternation *a ~ e* in the latter is conditioned by the back or front root vowels, respectively. The former suffix is identical to the 3rd person plural marker of the subjective conjugation. In absolute declension, both markers can appear (also with the same lexeme), and their distribution is not clear. One can only notice that in Plotnikova's corpus the former appears roughly twice more often. The latter, in turn, is the only one used in possessive declension as well as with demonstrative pronouns.

In absolute declension, case markers always follow the number markers, as shown in Table 5.

Table 5: Kamas nominal inflection: absolute declension (after Klumpp 2022a: Table 39.4, 825).

Case	Singular	Plural 1	Plural 2	Singular	Plural 1	Plural 2
	'river'	'rivers'	'rivers'	'rib'	'rib'	'rib'
Nominative	*t'ăga*	*t'ăga-jaʔ*	*t'ăga-zaŋ*	*koːt*	*koːd-ajaʔ*	*koːt-saŋ*
Genitive	*t'ăga-n*	*t'ăga-ja-n*	*t'ăga-zaŋ*	*koːd-ən*	*koːd-aja-n*	*koːt-saŋ*
Accusative	*t'ăga-m*	*t'ăga-ja-m*	*t'ăga-zaŋ-əm*	*koːd-əm*	*koːd-aja-m*	*koːt-saŋ-əm*
Lative	*t'ăga-nə*	*t'ăga-ja-nə*	*t'ăga-zaŋ-də*	*koːt-tə*	*koːd-aja-nə*	*koːt-saŋ-də*
Locative	*t'ăga-gən*	*t'ăga-ja-gən*	*t'ăga-zaŋ-gən*	*koːt-kən*	*koːd-aja-gən*	*koːt-saŋ-gən*
Ablative	*t'ăga-gəʔ*	*t'ăga-ja-gəʔ*	*t'ăga-zaŋ-gəʔ*	*koːt-kəʔ*	*koːd-aja-gəʔ*	*koːt-saŋ-gəʔ*
Instrumental	*t'ăga-z'əʔ*	*t'ăga-ja-z'əʔ*	*t'ăga-zaŋ-z'əʔ*	*koːt-s'əʔ*	*koːd-aja-z'əʔ*	*koːt-saŋ-z'əʔ*

Two processes should be mentioned here which operate in inflection and will be relevant for the further analysis, vowel epenthesis and intervocalic voicing. Both can be seen in forms of the noun *koːt* 'rib' whose stem, exceptionally for a Kamas noun, ends in an obstruent. First, an epenthetic vowel /ə/ appears between a consonantal root and the suffixes of Genitive *-n*, Accusative *-m*, and plural *-jaʔ ~ -iʔ* (the same goes for Plotnikova's *-ʔi ~ -ʔjə*). This happens not only after obstruents, which are mainly restricted to loans (either Turkic or Russian), but also after sonorants; cf. examples from Plotnikova's corpus: *ulaːr* 'sheep', *ulaːr-ən* 'sheep-GEN', *ulaːr-əʔi* 'sheep-PL'.

The second process only applies to voiceless obstruents: these become voiced in intervocalic position, namely before an epenthetic vowel. In addition to illustrations in Table 5, here is another example from Plotnikova's data: *d'irək* 'settler' (< Turkic), *d'irag-əʔi* 'settler-PL'. Conversely, the initial obstruent in suffixes is devoiced after a stem-final obstruent (cf. *koːt-tə* 'rib-LAT', *koːt-saŋ* 'rib-PL', etc. in Table 5).

In possessive declension, the order is generally (number-)case-possession, except that the Instrumental case suffix -z'əʔ comes last. Only four cases are distinguished: Nominative is merged with Accusative and Genitive (zero marker), and Lative with Locative. The Lative/Locative marker -gən and the Ablative -gət lose their initial -g and undergo contraction with polysyllabic vocalic stems like *tura* 'house' (Table 6).

Note that a form like *tura-zaŋ-də* is ambiguous between Lative plural (absolute declension) 'to the houses' and Nominative/Accusative/Genitive plural with 3rd pers. singular possessor 'his/her houses'.

Table 6: Kamas nominal inflection: possessive declension, 3rd person possessor (after Klumpp 2022a: Table 39.5–39.7, 825–826).

Case	Possessor number		
	Singular	**Dual**	**Plural**
Singular possessee	'his/her house'	'their (du.) house'	'their (pl.) house'
Nominative/Accusative/Genitive	*tura-t*	*tura-dəj*	*tura-dən*
Lative/Locative	*tura:n-də*	*tura:n-dəj*	*tura:n-dən*
Ablative	*tura:t-tə*	*tura:t-təj*	*tura:t-tən*
Instrumental	*tura-t-s'əʔ*	*tura-dəj-z'əʔ*	*tura-dən-z'əʔ*
Plural possessee	'his/her houses'	'their (du.) houses'	'their (pl.) houses'
Nominative/Accusative/Genitive	*tura-zaŋ-də*	*tura-zaŋ-dəj*	*tura-zaŋ-dən*
Lative/Locative	*tura-zaŋ-gən-də*	*tura-zaŋ-gən-dəj*	*tura-zaŋ-gən-dən*
Ablative	*tura-zaŋ-gət-tə*	*tura-zaŋ-gət-təj*	*tura-zaŋ-gət-tən*
Instrumental	*tura-zaŋ-də-z'əʔ*	*tura-zaŋ-dəj-z'əʔ*	*tura-zaŋ-dən-z'əʔ*

2.3 Kamas word prosody

The word prosody system in pre-shift Kamas was described by Donner (Joki 1944: 126–127) as follows. The primary stress (expressed by an 'expiratory accent') typically falls on the first syllable, often with a secondary stress on the third syllable. The primary stress may shift to the second syllable if it is closed or has a long vowel. In disyllabic words, stress may fall on either syllable. Under Turkic influence, prominence can also be assigned to the word-final syllable, not only in Turkic loans but also in native lexemes. Klumpp (2022: 823) further notes: "The Turkic ultima accent, however, is a pitch accent, and we don't know how much it co-occurred with the previous patterns" (i.e. with the probably inherited Kamas primary stress patterns).

As for post-shift Kamas, the observations summarized below should be regarded as preliminary, still they will be important for the following discussion of the plural formation. A more thorough analysis of the word prosody in post-shift Kamas is a large topic deserving investigation on its own and cannot not be undertaken within the limits of this paper.

In Plotnikova's Kamas, I will distinguish between vowel lengthening and word 'accent.' Vowel lengthening (as opposed to phonemic length) can be the result of a process accompanying suffixation, or an exponent of the original stress in Russian loans (see §3.1 below on Russian stress). I will designate as accent the prominence of a syllable in the word expressed primarily by a tonal contour (e.g. falling or rising tone). I assume that the type of the tonal contour is imposed by the higher-level (i.e. phrase-level) prosody and features such as illocution type, information structure, etc. At the same time, the placement of the accent in a word is determined by its phonological and morphological structure. The accented syllable thus serves as anchor for tonal contours determined word-externally. A word can also be unaccented, depending on the utterance prosody. On the other hand, word accent can optionally be accompanied (reinforced) by lengthening. In the absence of long vowels or lengthening, vowel duration does not vary systematically between syllables (like in Russian). As for possible intensity effects, they do not play a leading role in Standard Russian stress, and so far I have not found any significant intensity patterns in Kamas either.

Contrary to Donner's description, Kamas accent in Plotnikova's data is not at all common on the first syllable, which might be explained by the tendency imposed by contact with Turkic having generalized since the 1910s. In uninflected words, the accent falls typically on the last syllable (*nüké* 'wife', *tibí* 'man', *koŋgoró* 'bell'). If a long vowel is present, it is by default only distinguished by duration (*bü:z'é* 'husband', *oro:má* 'cream', *kö:bergán* 'onion'), often with a sustained pitch level.[10]

Kamas inflectional suffixes, in turn, differ in their accentual properties. Some of them are normally accented, like the plural *-ʔí* (both as nominal plural and as verbal 3rd person plural agreement marker). It is always accented when no other suffixes follow, as in as in *bü:z'e-ʔí* 'man-PL'. When followed by a case suffix, it keeps accent in roughly speaking half of occurrences (e.g. *pa-ʔí-nə ~ pa-ʔi-nə́* 'tree-PL-LAT'; *na:bə-ʔí-nə* 'duck-PL-LAT', but *ine-ʔi-nə́* 'horse-PL-LAT'; *ine-ʔí-zi ~ ine-ʔi-zí* 'horse-PL-LAT'). Note that the root-final vowel preceding *-ʔí* is not lengthened, unlike before case markers (see below).

The second plural marker *-zaŋ/-zeŋ* is only accented when used alone (*tibi-zéŋ* 'man-PL', *es-séŋ* 'child-PL', *ine-zéŋ* 'horse-PL'). When followed by a case or posses-

---

10 Throughout this paper, the Kamas accent will be marked with acute (´). Long vowels in Kamas will be marked with the length diacritic (:), whether due to phonemic length, or Kamas-specific lengthening, or lengthening corresponding to original Russian stress in loans. In examples from Russian, stress will be marked with acute. For example, cf. Kamas *balo:ta-ʔí* 'swamps' (< Rus. *bolóto* [bɐlótɐ]), *nali:mə-ʔí* 'burbots' (< Rus. *nal'ím* [nɐl'ím]).

sive suffix, it is the latter that takes accent (e.g. *ku-zaŋ-dź* 'ear-PL-POSS.3SG', *es-seŋ-dź* 'child-PL-POSS.3SG'), that is, accent is word-final in forms with *-zaŋ/-zeŋ*. Root-final vowel is lengthened only occasionally, mostly in some roots in *-a* (e.g. *kaga-zaŋ-dź* ~ *kaga:-zaŋ-dź* 'brother-PL-POSS.3SG').

Lative *-nə/-də* is as a rule unaccented when following a vocalic root ending in a non-high vowel (*turá:-nə* 'house-LAT', *kubá:-nə* 'skin-LAT', *kö:-nə* 'winter-LAT'; but *nüké:-nə* ~ *nüke:-nə́* 'woman-LAT'); the root-final vowel is subject to automatic lengthening. In other contexts, i.e. after a high-vowel root or a consonantal root, it is more often found accented than not (*bü-nź* 'water-LAT', *tibi-nź* 'man-LAT', *il-dź* 'people-LAT', but *kóŋ-də* 'chief-LAT').

Instrumental *-zi* and Locative *-gən* are similar to Lative, being usually unaccented after non-high vowels (lengthened) and accented otherwise (*pá:-zi* 'tree-INS', *pa-t-sí* 'tree-POSS.3SG-INS'; *bü-zí* 'water-INS', *tibi-zí* 'man-INS'). However, they bear accent more frequently even after non-high vowels (*turá:-gən* ~ *tura:-gə́n* 'house-LOC', *iné:-zi* ~ *ine:-zí* 'horse-INS').

See §4.3 for patterns found in Russian loans in Plotnikova's data.

2.4 Notes on the use of nominal plural forms

Plural forms of nouns are not obligatory when a plural referent is denoted. As in many Uralic languages, singular nouns are found instead in numeral phrases. However, this is a tendency rather than a strict rule: one example of a plural noun with *šide* 'two' is found in Donner's collection,[11] and in the post-shift corpus a larger share of numeral phrases involves plural nouns.

The proportion of plural forms is lowest for the numeral *šide* 'two' (7,8%) and increases for 'three' and 'four', and the more so for higher numerals (22,4%), see Table 7.

Table 7: Use of plural and singular forms with numerals in the post-shift Kamas corpus.

Numeral	Singular noun	%	Plural noun	%	Total
šide 'two'	95	92,2%	8	7,8%	103
nagur 'three', *teʔtə* 'four'	54	84,4%	10	15,6%	64
higher numerals	45	77,6%	13	22,4%	58

---

11 AIN_1912_Frogwoman_flk.001.

One and the same word can be found in both plural and singular form in numeral phrases, with and without possessive markers:

(1) *nagur   kaga*
     three   brother
     'three brothers' (PKZ_196X_su0217.PKZ.024)

(2) *šide   kaga-t   amno-bi-ʔi.*
     two   brother-POSS.3SG   live-PST-3PL
     'There lived two brothers.' (PKZ_196x_su0228.001)

(3) *šide   kaga-ʔi*
     two   brother-PL
     'two brothers' (PKZ_196x_su0228.011)

(4) *šide   kaga-zaŋ-də*
     two   brother-PL-POSS.3SG
     'his two brothers' (PKZ_196x_LittleGoat.022)

Both native items and borrowings can bear plural marking. Particular marking strategies found in borrowings will be discussed in detail in §5.

3 Preliminaries on Russian

In this section, I will outline some aspects of Russian phonology and morphology which are relevant for the analysis of Russian loans in Kamas. These concern primarily the vowel system and some consonantal features, stress, and the basic nominal declension types. For more details on Russian phonetics and morphology see e.g. (Jones & Ward 1969; Timberlake 2004).

3.1 Russian phonology and practical transcription

The consonant system of Russian features a pervasive contrast in palatalization. The voicing contrast in obstruents is neutralized word-finally and before other obstruents. There are two voiceless affricates, /c/ (non-palatalized hissing, IPA /ts/) and /č/ (palatalized hushing, IPA /tʃ/ or /tɕ/). Realizations of the palatal /j/ range from a fricative [j] before a stressed vowel to a weak semi-vowel [i̯] or even zero. See the full inventory (in the non-IPA transcription used hereafter) in Table 8.

Table 8: Standard Russian consonant system (after Yanushevskaya & Bunčić (2015)).

	labial		dental/alveolar		postalveolar		palatal	velar	
stops	p p'	b b'	t t'	d d'				k k'	g g'
fricatives	f f'	v v'	s s'	z z'	š š':	ž		x x'	
affricates			c		č'				
nasals		m m'	n n'						
laterals			l l'						
trills			r r'						
approximants							j		

The vowel system of Standard Russian includes five vowels which are distinguished under stress: /i e a o u/. Stressed /i/ is realized as a central [ɨ] after 'hard' (non-palatalized and non-palatal) consonants and as [i] otherwise. It is a matter of long-standing debate whether /ɨ/ should be considered a separate phoneme; here I assume it is not, however it has no effect on the phenomena discussed. The two views can be illustrated by (Avanesov 1956) and (Bondarko 1998), among countless other works.

The vowel system is tightly intertwined with that of word stress. The position of stress is lexically dependent and can be on any syllable in a word. Speaking paradigmatically, the position of stress is stable in some lexemes but variable in others (see §3.2).

It is generally accepted that the main acoustic manifestation of stress in Standard Russian is vowel duration along with vowel quality changes in unstressed vowels (Zlatoustova 1956; Bondarko 1977; Knyazev 2006; Yanushevskaya & Bunčić 2015). Many vowel contrasts are neutralized in unstressed positions; see e.g. Timberlake (2004) and Iosad (2012). In general, three types of positions with regard to stress are distinguished:
(i)   full vowels in stressed syllables (no reduction);
(ii)   moderate reduction: immediate pretonic syllables and unstressed word-initial (onsetless) position; also possible in unstressed phrase-final open syllables;
(iii)   radical reduction: all other unstressed positions (pre- and post-tonic); possible in unstressed phrase-final open syllables.

Apart from those positions, neutralizations depend on the preceding consonant, if any, being palatalized (or palatal /j/) or not. For instance, in moderate reduction /a/ and /o/ merge as [ɐ] after non-palatalized consonants (cf. (5a–d)), but after palatalized ones /a/, /o/ and /e/ merge as [ɪ] (5e–h):

(5) a. /pastúx/   [pɐstúx]   'shepherd'
    b. /pastux'í/   [pəstʊx'í]   'shepherds'
    c. /solov'éj/   [səlɐv'éj]   'nightingale'
    d. /solov'jí/   [səlɐv'jí]   'nightingales'
    e. /č'as/   [č'as]   'hour'
    f. /č'así/   [č'ɪsɨ́]   'clock; hours'
    g. /š':ot/   [š':ot]   'account'
    h. /š':otá/   [š':ɪtá]   'accounts'

The consonants /š ž c/, non-palatalized in Modern Standard Russian but palatalized historically, present a deviation from the general pattern. In moderate reduction, both /e/ and /o/ after these consonants reduce to [ɨ]. However, /o/ in recent borrowings also reduces to [ɐ], as after 'normal' non-palatalized consonants. For /a/, there is variation across lexemes between [ɨ] and [ɐ] as outcomes of reduction.

All vowels except /u/ neutralize in radical reduction. In the contemporary version of the pronunciation norm, all vowels except /u/ also neutralize in moderate reduction after palatalized consonants. This is one of the properties of what has been known as the 'newer norm,' first described in 1950s–1970s. It is opposed to the 'older norm' where /i/ is kept distinct from /a o e/ in this context: e.g., the first vowel in l'isá [l'isá] 'fox' is distinct from that in l'esá [l'ɪsá] 'woods', n'osú [n'ɪsú] 'I am bringing' and v'ažú [v'ɪžú] 'I am knitting'. See Avanesov (1956) and Comrie, Stone & Polinsky (1996) for an account of pronunciation standards. Note that the transition from the 'older norm' to the 'newer norm' is a gradual process, and the two norms may to some extent coexist until now in the same speaker, depending on e.g. communication settings and speech tempo.

Plotnikova's pronunciation in Russian does not have the mentioned /i/ – /a o e/ merger and is in general very close to the 'older norm' with respect to vowels. I will thus take the 'older norm' neutralization patterns as the basis of transcriptions of the Russian forms throughout this paper; they are presented in Table 9. The symbols are adapted from those used in Iosad (2012) with some modifications. On the one hand, in moderate reduction I will distinguish [i], [ɨ] from [ɪ], [ɨ] to account for the absence of the merger. On the other hand, I will disregard finer differences between [i], [ɨ] under stress vs. in moderate reduction, as well as between [ɪ], [ʊ] in moderate vs. radical reduction.

In the present paper, Russian examples will be given in the non-IPA transcription presented in the inventories above. To summarize deviations from the IPA:
-   The following symbols are used for sibilants: s š š': z ž c č (cf. the corresponding IPA symbols: s ʂ ɕ: z ʐ ts tɕ).
-   The palatal /j/ is transcribed as [ɪ̯] intervocalically before non-stressed vowels and as [j] otherwise.

- Consonant palatalization is marked with an apostrophe: /t' k'/.
- The stressed vowel is marked with an acute: /á ó/. Stress is not shown in monosyllabic words.

Table 9: Vowel neutralizations in Standard Russian (based on Iosad 2012).

Position	Tonic		Moderate reduction (immediate pretonic)			Radical reduction	
	initial or after C'	after C incl. /š ž c/	initial or after C'	after /š ž c/	after C	after C'	after C incl. /š ž c/
Vowel	i	ɨ	i	ɨ		ɪ	ə
	e		ɪ	ɨ		ɪ	ə
	a		ɪ	ɨ(e)	ᵉ	ɪ	ə
	o		ɪ	ɨ(e)	ᵉ	ɪ	ə
	u			ʊ			ʊ

3.2 Russian nominal morphology

Russian nouns are classified into three genders – masculine, feminine and neuter – with a further distinction between animate and inanimate. Nouns inflect for number (singular and plural) and case. Six main cases are recognized: Nominative, Accusative, Genitive, Dative, Instrumental and Prepositional.[12] Case and number are expressed by cumulative suffixes, forming three main inflectional classes traditionally called 1st, 2nd and 3rd declension, with complex syncretism patterns and several zero markers for particular feature combinations. These classes correlate with gender:

- 1st decl.: feminine or masculine, NOM.SG /-a/
- 2nd decl.: masculine, NOM.SG zero; neuter, NOM.SG /-o/
- 3rd decl.: feminine, NOM.SG zero

Animacy is relevant for the choice of the accusative form. In masculine nouns of the 2nd declension and in all nouns in plural, Accusative is identical to Nominative for inanimate nouns, and to Genitive for animate nouns.

Several stress patterns can be found in nouns, with the stress bearing always on the stem (as in /kúč'a/ 'heap'), always on the case-number suffix if not zero (as in /v'enók/ 'wreath'), or alternating between these options.

---

12 Several secondary case forms are distinguished in different analytical approaches, but they will not be relevant for this discussion.

Tables 10 and 11 show paradigms of the types most relevant for further discussion. The case forms are given in phonetic transcription which will also be used further in the article. Note that unstressed word-final vowels are transcribed assuming moderate reduction, as if they were phrase-final (e.g. individually cited words). This seems to reflect most adequately the distinctions kept in Plotnikova's pronunciation of Russian loans. In fluent Russian speech, radical reduction is now the norm in this position.

As for phonetic shape of the case forms, two points are to be mentioned. First, non-palatalized stem-final consonants except /š ž c/ become palatalized before the suffix /-e/ of Dative (1st decl.) and Prepositional (1st and 2nd declension). Furthermore, the velars /k g x/ also become palatalized before the suffix /-i/ of the Nominative plural (also Accusative plural for inanimate nouns), as in [v'ɪnk'-í] 'wreath-NOM/ACC.PL' (cf. the forms marked with * in Tables 10 and 11).

Second, some lexical stems contain a so-called fleeting vowel which only surfaces before a zero suffix and is absent in other forms (cf. the forms marked with ** in Tables 10 and 11).

An exhaustive account of the Standard Russian nominal inflection including stress patterns ('accentual paradigms') is given in Zaliznyak (1967), to which the reader is referred for more details.

Table 10: Russian inflectional paradigms. Feminine nouns, 1st and 3rd declension.

	1st decl.	/s'ostra/	1st decl.	/kuč'a/	3rd decl.	/t'en'/
	feminine	'sister'	feminine	'heap'	feminine	'shadow'
	Singular	Plural	Singular	Plural	Singular	Plural
Nom.	s'ɪstr-á	s'óstr-ɪ̸	kúč'-ɐ	kúč'-i	t'én'	t'én'-i
Acc.	s'ɪstr-ú	s'ɪst'ór**	kúč'-ʊ	kúč'-i	t'én'	t'én'-i
Gen.	s'ɪstr-í	s'ɪst'ór**	kúč'-i	kúč'	t'én'-i	t'ɪn'-éj
Dat.	s'ɪstr-é*	s'óstr-əm	kúč'-ɪ	kúč'-ɪm	t'én'-i	t'ɪn'-ám
Instr.	s'ɪstr-ój	s'óstr-əm'i	kúč'-ɪj	kúč'-ɪm'i	t'én'-jʊ	t'ɪn'-ám'i
Prep.	s'ɪstr'-é*	s'óstr-əx	kúč'-ɪ	kúč'-ɪx	t'én'-i	t'ɪn'-áx

Note: * marks stem-final palatalization. ** marks fleeting vowel.

Table 11: Russian inflectional paradigms. Masculine and neuter nouns, 2nd declension.

	2nd decl.	/n'em'ec/	2nd decl.	/v'enok/	2nd decl.	/boloto/
	masculine	'German'	masculine	'wreath'	neuter	'swamp'
	Singular	Plural	Singular	Plural	Singular	Plural
Nom.	n'ém'ɪc	n'émc-ɪ̸	v'ɪnók**	v'ɪnk-í*	belót-ɐ /-o/	belót-ɐ /-a/
Acc.	n'émc-ɐ	n'émc-əf	v'ɪnók**	v'ɪnk-í*	belót-ɐ /-o/	belót-ɐ /-a/
Gen.	n'émc-ɐ	n'émc-əf	v'ɪnk-á	v'ɪnk-óf	belót-ɐ /-a/	belót

Table 11 (continued)

	2nd decl.	/n'em'ec/	2nd decl.	/v'enok/	2nd decl.	/boloto/
	masculine	'German'	masculine	'wreath'	neuter	'swamp'
	Singular	Plural	Singular	Plural	Singular	Plural
Dat.	n'émc-ʊ	n'émc-əm	v'ɪnk-ú	v'ɪnk-ám	bɛlót-ʊ	bɛlót-əm
Instr.	n'émc-əm	n'émc-əm'i	v'ɪnk-óm	v'ɪnk-ám'i	bɛlót-əm	bɛlót-əm'i
Prep.	n'émc-ɨ	n'émc-əx	v'ɪnk'-é*	v'ɪnk-áx	bɛlót'-ɪ*	bɛlót-əx

Note: * marks stem-final palatalization. ** marks fleeting vowel.

4 Plotnikova's Russian and Russian borrowings

4.1 Russian dialects in European Russia and in Siberia

Dialects of the European part of Russia are traditionally classified in three large groups: Northern, Southern and Central (Zaxarova & Orlova 2004). The latter group is more internally diverse and acts in some respects as a transitional zone between the more internally homogeneous Northern and Southern groups. Standard Russian originates from the Central dialect group.

One of the most salient dialectal parameters is the vowel behaviour in unstressed position. While many vowel oppositions are neutralized in Standard Russian as well as in Southern varieties, most Northern varieties preserve them at least in the immediate pretonic position. Thus, the root vowels in pretonic position in *vod-á* 'water-NOM.SG' vs. *trav-á* 'grass-NOM.SG' are kept distinct in the North ([vodá] vs. [travá]) but are merged in the South as well as in Standard Russian ([vɐdá], [trɐvá]). Some Northern and Southern dialects may also additionally distinguish /e/ vs. /ɛ/ or /e o/ vs. /ɛ ɔ/. Some other notable features of the Southern dialects include, for instance, the fricative realization of the voiced velar /g/ as [ɣ] (and [x] when devoiced) and the palatalized /-t'/ in the verbal 3rd person suffixes (e.g. /l'et'ít'/ '(it) is flying', vs. Std. Rus. /l'et'ít/).

Russian dialects in Siberia, in turn, are a late formation driven by several waves of migration from the European part of Russia and other territories. The variety spoken by Klavdiya Plotnikova belongs to the so-called 'old-dwellers' dialects' (Rus. *starožil'českie govory*; for the term, see Blinova (1971) and Gusev (2020)). They are grounded in the Northern Russian varieties but cannot be identified with them fully, since many features come from Southern and Central branches brought in by later migrations. In particular, many old-dwellers' dialects do have vowel neutralization similar to the Southern ones, but keep plosive [g] as the Northern ones do. Crucially, the vocalic system of Plotnikova's Russian is virtually the same as in Standard Russian.

Most Russian forms cited in this paper are those of Modern Standard Russian; specific dialectal forms are marked as 'dial.' (note that it is not always possible to pinpoint the exact dialectal provenance of such forms).

4.2 Characteristics of Plotnikova's Russian

Let us cite some key characteristics of Klavdiya Plotnikova's Russian variety. Some of them are in line with the Standard Russian:

- unstressed /a/ and /o/ are neutralized in [ɐ] after non-palatalized consonants (feature known in Russian dialectology as *akanye*), as in Std. Rus. The whole pattern of vowel neutralization is most close to the 'older norm' of Standard Russian
- the voiced velar /g/ is realized as a plosive [g] and devoiced into [k], as in Std. Rus. (not as a fricative [ɣ] devoiced into [x], as in Southern Russian dialects)
- the 3rd person suffixes end in a non-palatalized /-t/, as in Std. Rus.

There are also markedly non-standard features:

- word-final /t/ is lost in /-st/ clusters: [l'is], Std. Rus. [l'ist] 'leaf'
- the long hushing fricative, corresponding to Std. Rus. /š':/, is non-palatalized /š:/: [iš:ó], Std. Rus. [jɪš':ó] 'still; yet'
- the intervocalic /j/ is completely lost in suffixes, causing vowel contraction in verbs and adjectives: [rɐbótət], Std. Rus. [rɐbótəjɪt] '(s/he) works'; [v'érʊš:ɨ], Std. Rus. [v'érʊi̯ʊš':ɪj] 'believer'; [tɐká krɐs'ívɐ], Std. Rus. [tɐkáɪ̯ə krɐs'ívəi̯ə] 'so beautiful (fem.)'
- the hissing affricate is usually simplified into /s/: [sélaj], Std. Rus. [célɨj] 'whole'; [l'itófsɨ], Std. Rus. [l'itófcɨ] 'Lithuanians'
- suffix [-aj] in some forms instead of unstressed [-ɨj] in adjective declension masculine, cf. [sélaj], Std. Rus. [célɨj] 'whole'

Note that many non-standard features in Plotnikova's Russian are to some extent variable and alternate with their standard counterparts. Conversely, at rare occasions she also produces forms typical for Northern dialects such as those with non-neutralized pre-tonic /o/, e.g. [kɔsá] 'braid' (Std. Rus. *kosá* [kɐsá]), [kɔl'só] 'ring' (Std. Rus. *kol'có* [kɐl'có]), but these are apparently exceptions.

4.3 Stress patterns in Russian borrowings in Kamas

Recall that vowel duration is the main, although not the only, acoustic correlate of stress in Standard Russian. In Plotnikova's data, Russian loans typically exhibit considerable vowel lengthening at the original stress location. To a Russian ear, it sounds as if the original stress were preserved. This lengthening is largely independent of the Kamas accent which can be superimposed when some Kamas material is added to the Russian stem. A first impression is then that there are two stressed syllables in a word. However, these are two different kinds of prominence, and it is not a matter of difference in degree (primary vs. secondary stress).

My preliminary observations show that the original Russian stress is mainly reflected in Kamas in vowel duration, often keeping a sustained pitch level. At the same time, Kamas accent is mainly expressed as a tonal contour, i.e. involves some change in pitch level. Note that in Standard Russian, these two kinds of prominence most commonly coincide in one syllable, i.e. it is the stressed syllable which has longer duration, full vowel (non-neutralized) quality, and also bears phrasal tonal movements (Knyazev 2006).

As a further complication, some Kamas suffixes may attract phrasal tonal movements not on themselves, but on adjacent syllables (see §2.2); e.g. *nüké:-nə ~ nüke:-nə́* 'woman-LAT'. When attached to a Russian borrowing, the accent can thus fall on the stem and coincide with the original stress (6a) or not (6b-c):

(6) a. *st'ená:-nə* 'wall-LAT' < Rus. /st'ená/ [st'má]
     b. *sva:d'bá:-nə* 'wedding- LAT' < Rus. /svád'ba/ [svád'bɐ]
     c. *ka:rtač'ká:-nə* 'card-LAT' < Rus. /kártoč'ka/ [kártəč'kɐ]

5 Plural marking on borrowed Russian nouns

In the following sections, we will discuss the plural-forming strategies that are found with Russian loans in Kamas. Let us state from the beginning that there is no functional difference between those strategies; we will only be concerned with the analysis of the formal aspect. I will use the term 'stem' to denote all the Russian material to which Kamas elements are added; it is generally an entire wordform including a case-number suffix.

In most common cases, a Kamas plural suffix is added either to the Russian citation form, i.e. Nominative singular (§5.1), or to the Russian plural (§5.2). Alternatively, Russian plural can be used with no further Kamas marking (§5.3). There are also cases where the underlying Russian form is difficult to identify (§5.4). It may be ambiguous between Nominative singular and plural or deviate from both. While a larger part of these cases is amenable to one or more of the three main strategies, a fourth one will be introduced in §5.4.3, featuring a generic oblique stem. All the observed strategies will be summarized in §5.5.

5.1 Strategy A: Russian Nom. sg. + Kamas PL

The first strategy takes a regular Kamas rendering of the Russian citation form (Nom. sg.) and adds a Kamas plural marker. In the simplest case, the underlying Russian form is virtually unchanged. If present, the vocalic nominative ending is preserved, cf. *ka:pl'a-ʔí* 'drop-PL' (Rus. *kápl'-a* [kápl'ɐ] 'drop'), *kni:ška-ʔí* 'book-PL' (Rus. *knížk-a* [kn'íškɐ] 'book').

Consonant-final stems with zero nominative usually get an epenthetic reduced vowel before the Kamas plural, as in *stul-uʔí* 'chair' (Rus. *stul*). (Recall that I will reflect the rounding of epenthetic /ə/ > [u] in transcription; see §2.1). There are, however, a few exceptions where no epenthesis happened, such as *bl'i:n-ʔí* (alongside the expected *bl'i:n-əʔí*) 'pancakes' (see Table 12).

Table 12: Forms with and without epenthesis after consonant-final stems.

No	Kamas				Russian		
	Singular | Plural	Ep.	Translation		Sing. | Plural	Phonetic	Gram.
(1)	—				kapkán	[kɐpkán]	m, 2
	kapka:n-**ə**ʔí	+	'traps'		kapkáni	[kɐpkánɪ̯]	
(2)	—				lábaz (dial.)	[lábəs]	m, 2
	la:baz-**ə**ʔí	+	'storehouses'		lábazi	[lábəzɪ̯]	
(3)	—				stul	[stul]	m, 2
	stu:l-**u**ʔí	+	'chairs'		stulja	[stul'jɪ]	
(4)	al'en'		'reindeer'		ol'én'	[ɐl'én']	m, 2
	al'e:n'-ʔí (~ ale:n'-əʔí)	–*	'reindeer (pl.)'		ol'én'i	[ɐl'én'i]	
(5)	—				bl'in	[bl'in]	m, 2
	bl'i:n-ʔí (~ bl'i:n-əʔí)	–*	'pancakes'		bl'iní	[bl'iní]	
(6)	—				oxótn'ik	[ɐxót'n'ik]	m, 2
	axo:tnik-ʔí	–*	'hunters'		oxótn'iki	[ɐxót'n'ɪk'i]	
(7)	—				p'ečát'	[p'ɪčát']	f, 3
	p'ič'a:t'-ʔí	–*	'stamps'		p'ečát'i	[p'ɪčát'i]	

Note: * marks forms where epenthesis is expected but did not happen.

Another regular change is the voicing of intervocalic obstruents at morpheme boundaries. In native words, the consonants found in the coda are usually sonorants or /ʔ/; *ko:t* 'rib' is one of rare native nouns ending in an oral stop. In loanwords root-final obstruents are common (Klumpp 2022: 823), cf. *kujnek* 'shirt' (< Turkic), *plot* 'raft' (< Rus.), *kazak* 'Russian' (< Rus. *kazak* 'Cossaque'). Intervocalically, i.e. before a vowel-initial suffix, fortis (voiceless) consonants alternate with lenis (voiced), as in *koːd-ən* 'rib-GEN', and may undergo spirantization and sometimes deletion, as in *kazaɣ-ən ~ kazaːn* 'Russian-GEN' (Klumpp 2022: 823).

In Plotnikova's data we regularly find voicing, cf. *kujneg-əʔi* 'shirt-PL' and Russian loans in Table 13. In *pastuːɣ-uʔjú* 'shepherd-PL' (< Rus. *pastux* 'shepherd'), the voicing of the velar fricative creates [ɣ]. While [ɣ] is a usual realization of /g/ in Southern Russian dialects, Plotnikova's old dwellers' Russian has a plosive [g], like Standard Russian. Intervocalically, Standard Russian only allows [ɣ] as an allophone of /g/ in several lexemes related to religious contexts or elevated style, like *bog* 'God', *blago* 'good, blessing', and in a few interjections (Knyazev & Pozharickaja 2011: 331). Some of these, e.g. Gen. *bo[ɣ]a* and Dat. *bo[ɣ]u* 'God', are in fact found in Plotnikova's Russian. Otherwise, she only has intervocalic [ɣ] in Kamas, either in loans as a result of /x/ voicing (Table 13: 3), or as a realization of /g/ before back vowels, as in *kaga* [kaɣa] 'brother', *nagur* [naɣur] 'three'.

The intervocalic Kamas voicing is in contrast with Russian where, on the contrary, final obstruents are regularly devoiced, but underlying voiceless remain invariably voiceless. The presence or absence of voicing in consonant-final roots of Russian origin can thus be telling about whether a certain Kamas form is based on a consonant-final form (e.g. Nom. sg. in 2nd and 3rd declension) or a vowel-final one (e.g. Nom. pl.). Note that underlying Russian voiced obstruents will always be voiced in Kamas intervocalically, whether the form is based on singular or plural stem (compare *p'iroːg-uʔi* 'cake-PL' with a hypothetical **p'irog'iː-ʔi* based on Nom. pl.).

The voicing predictably does not happen: (i) when no epenthetic vowel is added, as in *axotnik-ʔi* 'hunters' (see Table 12 above); (ii) when the final consonant is part of a cluster, as in *volk* 'wolf', pl. *voːlk-əʔi*. The voicing is also not found in four cases where it is expected to take place (these are marked with * in Table 13).

Different processes can thus serve as cues to disambiguate Nom. sg. and Nom. pl. consonant-final stems: palatalization of velars (in Nom. pl.), voicing before epenthetic vowel, and epenthetic /ə/ rounding after rounded vowels. Ambiguous cases will be treated in §5.4.1.

Palatalization is only considered for stem-final velars, voicing for underlying voiceless, and rounding for forms with epenthetic vowels after rounded vowels; otherwise *n/a* is used. * marks forms where voicing is expected but did not happen. ** marks dialectal lexemes/variants. All Russian nouns are 2nd decl., masculine.

Strategy A is quite common, being found in 164 of the 258 plural tokens (63.6%); this does not include the cases where the Nominative singular as underlying form is one of possible alternative analyses (see § 5.4.1).

Table 13: Intervocalic voicing in Kamas plurals based on Russian consonant-final stems.

No	Kamas Sing. | Plural	Pal.	Vc.	Rd.	Translation	Russian Sing. | Plural	Phonetic	Gram.
(1)	p'irók / p'iro:**g**-u?í	–	n/a	+	'cake' / 'cakes'	p'iróg / p'irog'í	[p'irók] / [p'ireg'í]	m, 2
(2)	v'inók / v'ino:**g**-u?í	–	+	+	'wreathe' / 'wreathes'	v'enók / v'enk'í	[v'ınók] / [v'ınk'í]	m, 2
(3)	pastúx / pastu:**ɣ**-u?jú	–	+	+	'shepherd' / 'shepherds'	pastúx / pastux'í	[pestúx] / [pəstʊx'í]	m, 2
(4)	plot / plo:**d**-u?jú	n/a	+	+	'raft' / 'rafts'	plot / plotí	[plot] / [pletí]	m, 2
(5)	sv'etók / sv'eto:**g**-u?í	–	+	+	'flower' / 'flowers'	cv'etók / cv'etí / cv'etk'í	[cv'ıtók] / [cv'ıtí] / [cv'ıtk'í]	m, 2
(6)	— / trav'ən'n'i:**g**-ə?í	–	+	n/a	'grass mats'**	trav'en'n'ík / trav'en'n'ik'í	[trəv'ın'n'ík] / [trəv'ın'n'ık'í]	m, 2
(7)	plat / pla:**d**-ə?í	n/a	+	n/a	'scarf' / 'scarves'	plat / pláti	[plat] / [plátį]	m, 2
(8)	— / karaga:**z**-ə?í	n/a	+	n/a	'Karagas people'	karagás / karagási	[kəregás] / [kəregásį]	m, 2
(9)	— / tu:jez'-i?í	n/a	+	n/a	'birchbark vessels'**	tújes[13] / tújes'i	[tújıs'] / [tújıs'i] (?)	m, 2
(10)	— / xa:r'**uz**-ə?jə́	n/a	+	–	'graylings'	xár'ius / xár'iusi	[xár'ıʊs] / [xár'ıʊsį]	m, 2
(11)	kazák / kaza:**k**-ə?í	–	–*	n/a	'Cossack' / 'Cossacks'	kazák / kazák'i (~ kazak'í)	[kezák] / [kezák'i] (~ [kəzek'í])	m, 2
(12)	— / plato:**k**-u?jú	–	–*	+	'scarves'	platók / platk'í	[pletók] / [pletk'í]	m, 2
(13)	— / klo:**p**-u?í	n/a	–*	+	'bedbugs'	klop / klopí	[klop] / [klepí]	m, 2
(14)	— / kro:**t**-u?jú	n/a	–*	+	'moles'	krot / krotí	[krot] / [kretí]	m, 2

---

13 The Standard Russian form is *tújes*. This word has many dialectal variants, with differing second vowel, stress location and plural formation (and sometimes gender). The final consonant can be underlyingly /s/, /s'/, /z/ or /z'/; see (SRNG 2012: 205–206, 215–218, 223, 312–313). However, the variants recorded with initial stress and a palatalized final consonant seem to have no plural like *tújez'i*, thus I assume Plotnikova's form to be based on Nom. sg. *tújes'* with subsequent voicing.

5.2 Strategy B: Russian Nom. pl. + Kamas PL

The second strategy adds a Kamas plural marker on top of the Russian plural form. It is found in a total of 34 tokens of 16 lexemes (13,2%); see Table 14.

One might expect that nouns which in Russian appear only or predominantly in plural would naturally tend to keep their plural as Kamas stem. Indeed, we find several occurrences of Russian *pluralia tantum* here: *s'e:n'i-ʔí* 'porch' (Rus. *s'én'i*), *varo:ta-ʔí* 'gate' (Rus. *voróta*). Unexpectedly, *no:žn'itsa-ʔí-nə* 'scissors (LAT)' has a stem-final [a] instead of [ɨ] (Rus. *nóžn'ici*, also pl. t.; the singular form **nóžn'ica* does not exist in Standard Russian); this might be driven by analogy to the feminine nouns with variable stress discussed in §5.4.2. Anyway, this form bears no Russian plural marking and should be classified as Strategy A; it is however given at the end of Table 14 (17) for completeness.

Table 14: Forms with double plural marking in Kamas.

No	Kamas		Count	Russian		
	Plural	Translation		Sing. | Plural	Phonetic	Gram.
(1)	č'isə-ʔí*	'hours'	3	č'as	[č'as]	m, 2
				č'así	[č'ɪsí]	
(2)	č'e:r(')v'i-ʔí	'worms'	2	č'erv'	[č'erf']	m, 2
	č'e:r'vɨ-ʔí-zɨʔ	'worms (INS)'	1	č'érv'i	[čérv'i]	
(3)	č'ulk'i:-ʔí	'stockings'	1	č'ulók	[č'ʊlók]	m, 2
				č'ulk'í	[č'ʊlk'í]	
(4)	gn'o:zda-ʔí	'nests'	1	gn'ozdó	[gn'ɪzdó]	n, 2
				gn'ózda	[gn'ózde]	
(5)	kal'o:sa-ʔí	'wheels'	1	kol'osó	[kəl'ɪsó]	n, 2
				kol'ósa	[kel'óse]	
(6)	ko:r'əšk'i-ʔí,	'roots'	3	kor'ešók	[kər'ɪšók]	m, 2
	ka:r'əšk'i-ʔí		1	kor'ešk'í	[kər'ɪšk'í]	
(7)	kazl'a:ta-ʔjí	'goatlings'	1	kozl'ónok	[kezl'ónək]	m, 2**
				kozl'áta	[kezl'áte]	
(8)	ko:rn'i-ʔí	'roots'	1	kór'en'	[kór'ɪn']	m, 2
				kórn'i	[kórn'i]	
(9)	o:gurcɨ-ʔí	'cucumbers'	1	ogur'éc	[egʊr'éc]	m, 2
				ogurcí	[egʊrcí]	
(10)	raga:-zaŋ-də̂	'horns (POSS.3SG)'	1	rog	[rok]	m, 2**
				rogá	[regá]	
(11)	r'e:mn'i-ʔí,	'belts'	1	r'em'én'	[r'ɪm'én']	m, 2
	r'imn'i-ʔí*		2	r'emn'í	[r'ɪmn'í]	

Table 14 (continued)

No	Kamas		Count	Russian		
	Plural	Translation		Sing. | Plural	Phonetic	Gram.
(12)	salav'ji-ʔí*	'nightingales'	1	solov'éj	[səlev'éj]	m, 2
				solov'jí	[səlev'jí]	
(13)	s'eːn'i-ʔí	'porch (pl. t.)'	1	—	—	(m), 2
				s'én'i	[s'én'i]	
(14)	snapɨ-ʔí*	'sheaves'	1	snop	[snop]	m, 2
				snopí	[snepʃ]	
(15)	varoːta-ʔí	'gate (pl. t.)'	7	—	—	(n), 2
				voróta	[veróte]	
(16)	veršk'i-ʔí*	'tops'	4	v'eršók	[v'ɪršók]	m, 2
				v'eršk'í	[v'ɪršk'í]	
(17)	noːžn'itsa-ʔí-nə	'scissors (LAT) (pl. t.)'	1	—	—	(f), 1
				nóžn'ici	[nóžn'ɪci]	

* marks Kamas forms where the original Russian stress is not manifested in vowel lengthening.
** *kozl'ónok* 'goatling' and *rog* 'horn' have irregular plurals.

At least for some other nouns in this group, the plural form is highly salient in Russian and perhaps more natural than the respective singular form. Thus, Rus. *čulk'í* 'stockings' and *rogá* 'horns' are pair nouns and usually appear in plural. The form *časí* is homonymous between the plural of 'hour' (the meaning which is referred to in the corpus by *č'isə-ʔi*) and 'clock, watch' which is a *pluralia tantum* noun. The words *v'eršk'i* 'tops' and *kor'ešk'i* 'roots' in the corpus come from a fairy tale which is traditionally known in Russian as *V'eršk'i i kor'ešk'i* ('Tops and roots'). In this tale they denote respectively the upper and lower parts of cultivated plants (rye and turnips) shared between the two protagonists and are used as mass nouns. (There is also an occurrence of the singular form *veršok* 'top' in another text). In addition to *kor'ešk'í* which in Russian is a diminutive derivation from *kóren'* 'root', the plural of the underived stem, *koːrn'i-ʔí* 'roots' is also found in the corpus.

Some observations are made concerning the manifestation of Russian stress in these forms. In the forms where the Russian stress is originally not word-final, such as e.g. (Table 14: 2, 4, 5), it is expressed by vowel lengthening as expected. In contrast, those forms where Russian stress is originally word-final behave differently. Only one of them (Table 14: 3) shows stem-final lengthening before the accented Kamas suffix *-ʔí*. The form *ragaː-zaŋ-də́* 'its horns' (Table 14: 11) also shows stem-final lengthening, but the following suffix is the unaccented plural *-zaŋ* (see §2.3). Five forms marked with * show no lengthening at all, like *č'isə-ʔí* 'hours' (Table 14: 1, 11, 12, 14, 16). Finally, the lengthening in (Table 14: 6, 9, 11) appears on the first syllable and not on the originally stressed one. Note that *gn'oːzda-ʔí* 'nests' and *kal'oːsa-ʔí* 'wheels' would have a stem-final stress in Nom. sg. (Rus. *gn'ozdó, koľosó*). We will return to these observations in §5.4.

Let us remark that such double Russian-Kamas marking (see e.g. Matras (2009: 174) on double marking of plural on loans) is not unique to the category of plural. A similar phenomenon is found in several examples involving instrumental case (in predicative function and expressing means of transportation).

5.3 Strategy C: Russian Nom. pl. only

The third strategy includes the most straightforward cases, i.e. those where the ready-made Russian plural form is taken to represent the plural in the Kamas text. They are not numerous; 16 tokens of 11 nouns (6,2%) are found in the corpus (see Table 15).

Table 15: Forms with Russian-only plural marking.

No	Kamas Plural	Pal.	Fl.v.	Translation	Count	Russian Sing. | Plural	Phonetic	Gram.
(1)	čʼəsʼí	n/a	n/a	'hours'	1	čʼas / čʼasí	[čʼas] / [čʼɪsʼí]	m, 2
(2)	čʼe:rvə	n/a	n/a	'worms'	1	čʼervʼ / čʼérvʼi	[čʼerfʼ] / [čʼérvʼi]	m, 2
(3)	já:blak'i	+	n/a	'apples'	3	jábloko / jáblok'i	[jábləke] / [jáblək'i]	n, 2
(4)	kanfé:ti	n/a	n/a	'candies'	1	konfʼéta / konfʼéti	[kenfʼéte] / [kenfʼétʲi]	f, 1
(5)	ka:rʼəšk'í	+	–	'roots'	2	korʼešók / korʼešk'í	[kərʼɪšók] / [kərʼɪšk'í]	m, 2
(6)	kú:čʼi	n/a	n/a	'heaps'	1	kúčʼa / kúčʼi	[kúčʼɪ] / [kúčʼi]	f, 1
(7)	rʼimnʼí	n/a	–	'belts'	1	rʼemʼénʼ / rʼemnʼí	[rʼɪmʼénʼ] / [rʼɪmnʼí]	m, 2
(8)	salavʼjí	n/a	–	'nightingales'	1	solovʼéj / solovʼjí	[səlevʼéj] / [səlevʼjí]	m, 2
(9)	xo:xlɨ	n/a	–	'Ukrainians'	1	xoxól (colloq.) / xoxlí	[xexól] / [xexlʼí]	m, 2
(10)	vʼiršk'í	+	–	'tops'	2	vʼeršók / vʼeršk'í	[vʼɪršók] / [vʼɪršk'í]	m, 2
(11)	žo:ludʼí	n/a	n/a	'acorns'	2	žólud' / žóludʼi	[žólʊtʼ] / [žólʊdʼi]	m, 2

These forms evidently lack any Kamas elements and can be readily identified as Russian plural (and not singular) forms: in addition to the final vowel representing the plural ending (/-i/), most of them demonstrate expected stem alternations in vowels (absence of the fleeting vowel, /a/ reduction) and consonants (palatalization of /k/).

Russian-only marked plurals are clearly a minority option; not only their number is low, but for eight of the nouns we also find plurals formed with an overt Kamas plural marker, i. e. with one of the other strategies (see Table 16). For instance, alongside *salav'jí* 'nightingale.PL' we also find *salav'ji-ʔí* 'nightingale.PL-PL'.

Table 16: Russian loans with concurrent plural forming strategies.

No	Kamas			Count	Russian		
	Plural	**Translation**	**Strategy**		**Sing. | Plural**	**Phonetic**	**Gram.**
(1)	jáblak'i	'apples'	C	2	jábloko	[jáblǝke]	n, 2
	ja:blaka-ʔí		A	1	jáblok'i	[jáblǝk'i]	
(2)	xoxlí	'Ukrainians'	C	1	xoxól (colloq.)	[xexól]	m, 2
	xaxo:l-uʔí		A	1	xoxlí	[xexlí]	
(3)	č'ǝsí	'hours'	C	1	č'as	[č'as]	m, 2
	č'isǝ-ʔí		B	3	č'así	[č'ɪsí]	
(4)	č'e:rvǝ	'worms'	C	1	č'erv'	[č'erf']	m, 2
	če:r(')v'i-ʔí,	'worms (INS)'	B	2	č'érv'i	[č'érv'i]	
	če:r'vi-ʔí-z'i		B	1			
(5)	kar'ǝšk'í	'roots'	C	2	kor'ešók	[kǝr'ɪšók]	m, 2
	ka:r'ǝšk'í-ʔí,		B	1	kor'ešk'í	[kǝr'ɪšk'í]	
	ko:r'ǝšk'í-ʔí		B	3			
(6)	r'emn'í	'belts'	C	1	r'em'én'	[r'ɪm'én']	m, 2
	r'e(:)mn'i-ʔí		B	3	r'emn'í	[r'ɪmn'í]	
(7)	salav'jí	'nightingales'	C	1	solov'éj	[sǝlev'éj]	m, 2
	salav'ji-ʔí		B	1	solov'jí	[sǝlev'jí]	
(8)	v'eršk'í	'tops'	C	2	v'eršók	[v'ɪršók]	m, 2
	v'eršk'i-ʔí		B	4	v'eršk'í	[v'ɪršk'í]	

5.4 Problematic cases: Various stems + Kamas PL

The remaining occurrences of plural in loans pose problems: while they evidently contain a Kamas plural marker, it is not clear to which stem this marker is attached. Some of them are just ambiguous between Russian singular and plural forms (§5.4.1). In more intricate cases (§5.4.2–5.4.3) the stem apparently coincides neither with the Russian singular nor with the Russian plural form. No single explanation seems to fit all the cases. Let us examine them in more detail.

5.4.1 Russian consonant-final nouns: Nom. sg./Nom. pl. ambiguity

Several consonant-final nouns with a zero NOM.SG suffix in Russian would require an epenthetic /ə/ if their singular form is taken as the Kamas stem, or take an unstressed /-i/ NOM.PL suffix ([ɪ / ɨ] in moderate reduction) if Russian plural is used. Most of them are masculine and one is feminine (see Table 17). Phonetically, it is hard to reliably distinguish between a Kamas /ə/ and a reduced Russian post-tonic /i/, especially given the high lability of the epenthetic schwa. While in some other cases singular and plural stems can be disambiguated e.g. thanks to the Russian stem alternations, Kamas intervocalic voicing or schwa rounding (cf. Table 13), for the forms listed in Table 17 both interpretations are in principle possible.

The stem-final vowel in [faši:stɨʔí] 'fascists' (Table 17: 2) is pronounced closer to [ɨ] than to [ə], however it follows a stressed [í] and could possibly be considered a realization of an assimilated Kamas /ə/ as well as of the Russian /i/ of the plural suffix. The two forms for 'soldiers' (Table 17: 11) lack intervocalic voicing, which would imply plural stem, but their stem-final vowels are [a] and [ə] instead of the expected [ɨ]. Finally, for the word of dialectal origin *mura:š-əʔí* 'ants' (Table 17: 7), the declension type and the plural form are not clear.

Table 17: Russian consonant-final root nouns: masculine and feminine.

No	Kamas		Count	Russian		
	Plural	Translation		Singular | Plural	Phonetic	Gram.
(1)	al'e:n'(-)ə(-)ʔí	'reindeer (pl.)'	3	ol'én'	[el'én]	m, 2
	al'e:n'(-)ə(-)ʔi:-z'í	(INS)	4	ol'én'i	[el'én'ɪ]	
(2)	faši:st(-)ɨ(-)ʔí	'fascists'	1	fašíst	[fešíst]	m, 2
				fašísti	[fešístɨ]	
(3)	kamba:jn(-)ə(-)ʔí	'harvester'	1	kambájn	[kembájn]	m, 2
				kambájni	[kembájnɨ]	
(4)	kapka:n(-)ə(-)ʔí	'traps'	1	kapkán	[kepkán]	m, 2
				kapkáni	[kepkánɨ]	
(5)	kla:d'(-)ə(-)ʔí	'stacks'	1	klad'	[klat']	f, 3
				klad'i	[kláд'ɪ]	
(6)	la:baz(-)ə(-)ʔí	'storehouses'	1	lábaz	[lábəs]	m, 2
				lábazi	[lábəzɨ]	
(7)	mura:š(-)ə(-)ʔí	'ants' (dial.)	1	muráš	[mʊráš]	?
				?	?	
(8)	nali:m(-)ə(-)ʔí	'burbots'	1	nal'ím	[nel'ím]	m, 2
				nal'ími	[nel'ímɨ]	

Table 17 (continued)

No	Kamas		Count	Russian		
	Plural	Translation		Singular | Plural	Phonetic	Gram.
(9)	padva:l(-)ə(-)ʔí-nə	'cellars (LAT)'	1	podvál	[pɐdvál]	m, 2
				podváli	[pɐdválʲ]	
(10)	pʼilʼmʼeːnʼ(-)ə(-)ʔí	'dumplings'	1	pʼelʼmʼénʼ	[pʼɪlʼmʼénʼ]	m, 2
				pʼelʼmʼénʼi	[pʼɪlʼmʼénʼɪ]	
(11)	solda:t(-)a(-)ʔí	'soldiers'	1	soldát	[sɐldát]	m, 2
	salda:t(-)ə(-)ʔí		1	soldáti	[sɐldátʲ]	
(12)	zvèrʼ(-)ə(-)ʔí	'beasts'	1	zvʼerʼ	[zvʼerʼ]	m, 2
				zvʼerʼi	[zvʼérʼɪ]	

The Russian nouns *bolóto* 'swamp' and *górlo* 'throat' are originally neuter and their Nom. pl. forms *bolóta* and *górla* are normally homophonous to the Nom. sg. form due to the merger of post-tonic /a/ and /o/. Finally, several words belonging morphologically to the adjective declension have homophonous (masculine) nominative singular and plural forms in Plotnikova's old dwellers' dialect, in contrast to Standard Russian. These are listed in Table 18.

I will classify this group of forms ambiguous between Nom. sg. and Nom. pl. stems as Strategy A/B.

Table 18: Russian consonant-final roots: neuter gender and morphological adjectives.

No	Kamas		Ct.	Russian (dial. / Std.)		
	Sing. | Plural	Translation		Singular | Plural	Phonetic	Gram.
(1)	balo:tá:-gən	'swamp (LOC)'	1	bolóto	[bɐlóte]	n, 2
	balo:ta-ʔí	'swamps'	1	bolóta	[bɐlóte]	
(2)	—	'throats' (POSS.3SG)	1	górlo	[górle]	n, 2
	go:rla:-zaŋ-dǝ			górla	[górle]	
(3)	—	'copper (adj. pl.)'	1	mʼédni / mʼédnij	[mʼédnʲ] / [mʼédnəj]	adj (m)
	mʼeːdnə-ʔí			mʼédni / mʼédnije	[mʼédnʲ] / [mʼédnəɪɪ]	
(4)	—		1	soxáti / soxátij	[sɐxátʲ] / [sɐxátəj]	adj (m)
	saxa:tə-ʔí	'elks'		soxáti / soxátije	[sɐxátʲ] / [sɐxátəɪɪ]	
(5)	—		1	vʼerušʼi / vʼerujušʼij	[vʼérušʼɪ] / [vʼérujʊšʼij]	adj (m)
	vʼeːrušə-ʔí	'believers'		vʼerušʼi / vʼerujušʼije	[vʼérušʼɪ] / [vʼérujʊšʼɪɪ]	
(6)	—		1	zʼelʼóni / zʼelʼónij	[zʼɪlʼónʲ] / [zʼɪlʼónəj]	adj (m)
	zʼelʼo:na-ʔí	'green (pl.)		zʼelʼóni / zʼelʼónije	[zʼɪlʼónʲ] / [zʼɪlʼónəɪɪ]	

5.4.2 Russian feminine nouns with variable stress: Nom. sg. with stress shift in Kamas

Another group of less trivial cases includes feminine nouns with variable stress (Table 19: 1–6). The Kamas plural forms differ from Russian Nom. sg. in stress placement (which in the latter falls on the NOM.SG suffix) and in several cases also in the root vowel. At the same time, they differ from the Nom. pl. forms in their final vowel, which should otherwise be [-ɨ].

Table 19: Russian feminine nouns with variable stress.

No	Kamas		Count	Russian		
	Singular | Plural	Translation		Sing. | Pl.	Phonetic	Gram.
(1)	—		1	pč'olá	[pč'ɪlá]	f, 1
	č'o:la-ʔí*	'bees'		pč'óli	[pč'ólɨ]	
(2)	—		2	grozá	[grezá]	f, 1
	gro:za-ʔí	'thunderstorms'		grózi	[grózɨ]	
(3)	kapná:-nə (LAT)	'haystack'	1	kopná	[kepná]	f, 1
	ko:pna-ʔi	'haystacks'	2	kópni	[kópnɨ]	
(4)	st'əná:-nə (LAT)	'wall'	2	st'ená	[st'ɪná]	f, 1
	st'e:na-ʔí	'walls'	1	st'éni	[st'énɨ]	
(5)	str'əlá; str'əlá:-zi (INS)	'arrow'	2; 2	str'elá	[str'ɪlá]	f, 1
	str'e:la-ʔí	'arrows'	1	str'éli	[str'élɨ]	
(6)	—		1	zv'ozdá	[zv'ɪzdá]	f, 1
	zv'o:zda-ʔí	'stars'		zv'ózdi	[zv'ózdɨ]	
(7)	—		1	sosná	[sesná]	f, 1
	sasna-ʔí	'pines'		sósni	[sósnɨ]	
(8)	s'estrá; s'estrá-m (1SG), etc.	'sister'	2; 10	s'ostrá	[s'ɪstrá]	f, 1
	s'estra:-zaŋ-də́	'her sisters'	2	s'óstri	[s'óstrɨ]	

* The initial cluster *pč'-* is simplified in Kamas.

The questions we will address now are: why regular forms based on Nom. sg. or Nom. pl. stems are not used for these nouns, and next, why the particular 'hybrid' forms in Table 19 are chosen instead.

Let us first compare hypothetical alternative forms (hereafter marked with *) based on Nom. sg. and Nom. pl. stems:

(7) a. **č'ela:-ʔí* 'bee-PL' < Rus. *pč'elá* [pč'ɪlá] (Nom. sg.)

     b. **č'o:lɨ-ʔí* 'bee.PL-PL' < Rus. *pč'óli* [pč'ólɨ] (Nom. pl.)

Since there is no apparent difference in morphological structure between forms in (7ab) and those listed in §5.1 (Strategy A) and §5.2 (Strategy B) respectively, I suppose that the reasons to disprefer the two stems relate to word prosody and phonotactics.

The hypothetical form in (7a) has lengthening (reflecting the original Russian stress) on the syllable directly preceding the plural suffix -ʔí. In Plotnikova's data, this suffix is typically accented, being prominent both with respect to vowel duration and tonal change (see §2.3). However, it does not trigger the automatic lengthening of preceding vowels, as other consonant-initial suffixes do. Furthermore, it appears that lengthening corresponding to original Russian stress in loans almost never appears before -ʔí. Among all the tokens in our data there is probably only one example where the original Russian word-final stress is preserved and directly followed by -ʔí: čulk'i-ʔí 'stockings' (see Table 14: 3).

I hypothesize that it is the absence of automatic lengthening before the suffix -ʔí in Kamas words which extends to Russian loans and disfavours forms with stem-final lengthening like (7a). It remains yet to understand what forms are produced instead, and why (7b) is also dispreferred.

The last two entries in Table 19 feature feminine nouns of the same declension type, which behave differently from those discussed above. One of them, *sasna-ʔí* 'pines', has no lengthening on the stem syllables (i.e. no overt realization of the original Russian stress), looking similar to many native Kamas words (e.g. *jama-ʔí* 'boots'). Recall other forms with unexpressed original stress in Table 14 (Strategy B), such as *č'isə-ʔí* 'hours'. The conflict of Russian stress and non-lengthening is resolved here by omitting the former.

The last form, *s'estra:-zaŋ-də́* 'her sisters',[14] is a possessive plural and thus necessarily uses the plural suffix *-zaŋ*, unaccented before a possessive suffix. In contrast to the plural *-ʔí*, the suffix *-zaŋ* does trigger lengthening of non-high vowels (especially of *-a*) in Kamas items, although not very regularly (see §2.3). The presence of Russian-stress lengthening before *-zaŋ* is thus unproblematic; cf. also *raga:-zaŋ-də́* 'its horns' (Table 14: 11).

Interestingly, when the Russian stress is not stem-final, the suffix *-zaŋ* can still trigger stem-final vowel lengthening, as in *go:rla:-zaŋ-də́* 'their throats' (lit. 'his/her throats') (Table 18: 2) and *žerno:fka:-zaŋ-də́* 'his hand mill' (Table 20: 6) with lengthening (of different origin) on two stem vowels each. Such double lengthening also occurs in case-marked singular forms, where the stem-final vowel also receives Kamas accent, e.g. *ka:rtač'ká:-nə* 'card-LAT' (see Table 20).

---

14 One of the two occurrences of this form apparently has no stress at all due to the weak phrasal position.

Note that the stem-final vowel, despite its increased duration, appears as [aː] even when corresponding to an underlying /o/ in Russian, i.e. the neutralizing effect of Russian reduction is preserved. This is evident in singular forms of the originally neuter nouns ending in /-o/ in Nom. Sg., *baloːtáː-gən* 'swamp (LOC)' and *kariːtáː-z'i?* 'trough (INS)' (Table 20: 4–5). This is in contrast to what happens in the discussed feminine forms from Table 19.

Table 20: Stem-internal lengthening and stem-final accent in Russian loan nouns.

No	Kamas		Russian		
	Wordform	Translation	Sing. | Plural	Phonetic	Gram.
(1)	ziːpkáː-gən	'cradle-LOC'	zíbka (dial.) zíbk'i	[zípke] [zípk'i]	f, 1
(2)	svaːd'báː-nə	'wedding-LAT'	svád'ba svád'bi	[svád'be] [svád'bi̦]	f, 1
(3)	kaːrtač'káː-nə	'card-LAT'	kártoč'ka kártoč'k'i	[kártəč'ke] [kártəč'k'i]	f, 1
(4)	baloːtáː-gən	'swamp-LOC'	bolóto bolóta	[beˡóte] [beˡóte]	n, 2
(5)	kariːtáː-z'i?	'trough-INS'	koríto koríta	[keríte] [keríte]	n, 2
(6)	žernoːfkaː-zaŋ-də́	'hand.mill-PL-3SG'	žernóvka (dial.) žernóvk'i	[ž̦ı̦rnófke] [ž̦ı̦rnófk'i]	f, 1

As concerns the difference between the hypothetical plural-based form (7b) and the observed ones, they contrast in their stem-final vowel. The observed Kamas forms have a stem-final [a], as in *č'oːla-?í* 'bees'. The Russian plural in these words ends in unstressed /i/ following a non-palatalized consonant, yielding [ı̦] in moderate reduction: *pč'óli* [pč'ólı̦] 'bees'; this would give a stem-final /-i/ in Kamas. Two factors can be considered: (i) plural stems in general are dispreferred, (ii) plural stems with stem-final /-i/ are dispreferred as the base for plurals. Or, formulated positively, (i) singular stems are preferred, (ii) stem-final /-a/ is preferred.

The first factor (i) is probably the main motivation. Unambiguous plural stems (i.e. Strategy B) are by themselves rare among the plural tokens discussed, and roughly half of them are motivated either by higher naturalness of the plural form (*pluralia tantum*, pair nouns, etc.) or by avoiding stem-final stress. The number of remaining cases, for which no direct motivation is evident, is still smaller.

There is also some secondary evidence in support of (ii). As a systemic consideration, recall that there is no native /i/ in Kamas. My hypothesis is that –perhaps for this reason– when choosing between the Nom. sg. and Nom. pl. stems to construct a plural form, stem ending in /-ɨ/ are chosen by Plotnikova even more seldom than other plural stems. For the nouns in the corpus these are predominantly stems in /-i/ and a few ones in /-a/.

Consider, for instance, two feminine nouns which would be expected to use their plural stem: *noːžnʼitsa-ʔí-nə* 'scissors' (Rus. pl. t. *nóžnʼici* [-ɨ]; Table 14: 17) and *lɨːža-ʔí* 'ski-PL' (Rus. *líža*, Nom. pl. *líži* [-ɨ]). The former is a *pluralia tantum* but the stem used by Plotnikova is in fact a (non-existent) singular stem in /-a/. The second is a pair noun, and again uses the Nom. sg. stem in /-a/. Of the four *pluralia tantum* and pair nouns which do use their Nom. pl. stem, two have stems in /-a/ and two others in /-i/ (Table 14: 3, 10, 13, 15).

Plurals with stem-final /-ɨ/ do occur in the corpus, e.g. *čʼeːrʼvɨ-ʔɨ-ziʔ* 'worms (INS)', *oːgurcɨ-ʔí* 'cucumbers' (Table 14: 2, 9), *kanfʼeːtɨ* 'candies' (Table 15: 4) but rarely. We find a total of 12 such tokens (excluding instances of Strategy A/B ambiguous between the Nom. sg. and Nom. pl. stems). Compare this to 89 tokens of nouns whose Nom. pl. stem also ends in /-ɨ/ but for which another stem is chosen for the Kamas plural (most often Nom. sg., but also special variants discussed in §5.4.2–5.4.3). This ratio is more than twice smaller than that for stems in /-i/ (Table 21); the number of stems in /-a/ is altogether very small, but the ratio for them is still higher.

Table 21: Proportion of Nom. pl. stems to other stems depending on the stem-final vowel.

Nom. pl. stem-final vowel	Tokens based on Nom. pl.	Tokens based on other stems	Ratio (Nom. pl. to other stems)
/-ɨ/	12	89	0.135
/-i/	27	90	0.3
/-a/	4	4	1.0

In sum, it looks like the Nom. pl. Russian stems, and especially those with stem-final /-ɨ/, are dispreferred in Plotnikova's Kamas. Note that in Plotnikova's Russian, this context is well represented and unproblematic. For instance, alongside Kamas *stʼeːna-ʔí* 'walls' with stem-final [a] we find the regular Nom. pl. *stʼénɨ* in the Russian fragment of the same sentence:

(8) p'eš-tə   kămnə-bia-m   bar   **stʼeːna-ʔi**   büː-zi...
    stove-LAT   pour-PST-1SG   PTCL   wall-PL   water-INS
    [Rus.:]   *oblʼilá*   **stʼénɨ**   étʼi
                pour.over.PST   wall.PL   these
    '[She] poured [water] on the oven, poured water all over the walls... [Rus.:] poured over these walls.'
    (PKZ_196x_SU0218.023)

This leaves the question open: what is the nature of the stem-final [a] which appears in the six nouns in Table 19? The unstressed /ɨ/ in non-palatalized context is normally realized as [ɨ] (stem-finally) or possibly as [ə], but not as a low vowel [a]. An alternative would be to consider [a] a realization of the Kamas epenthetic /ə/ directly attached to the Russian root (with no Russian vocalic case-number suffix). However, given the long [o:] in 4 of the 6 nouns in Table 19, a /ə/ attached to the Russian root would have yielded [u] in those nouns (see §2.1), which is however not the case. Cf. the hypothetical plural of 'bee' in (9a) to the attested form with a stressed /o/ in (9b):

(9) a. *č'o:l-[u]ʔí 'bee.PL-PL' < č'o:l+ə+ʔí < Rus. pč'ól- (pl. stem)

     b. xaxol-[u]ʔí 'Ukrainian-PL' < xaxo:l+ə+ʔí < Rus. dial. xoxól   (sg. stem)

We must then recognize that the stem-final [a] is distinct from both regular /ɨ/ and /ə/. The best solution seems to admit a Russian Nom. sg. stem altered by a stress shift to the first syllable. Apart from just shifting the stress (i.e. lengthening) to the left, the vowel taking stress is restored to its underlying quality, thereby undoing the effects of Russian reduction: [o:] in (Table 19: 1–3, 6) and [e:] in (Table 19: 4, 5). While this seems to be quite a costly operation, we do find evidence of such correspondences in other contexts, repeated here from Tables 14–15.

Table 22: Stress shift with and without vowel restoration.

No	Kamas		Count	Russian		
	Plural	Translation		Sing. | Plural	Phonetic	Gram.
(1)	ko:rn'i-ʔí	'roots'	1	kór'en' kórn'i	[kór'ɪn'] [kórn'i]	m, 2
(2)	kar'əšk'í	'roots'	2	kor'ešók kor'ešk'í	[kər'ɪšók] [kər'ɪšk'í]	m, 2
(3)	ka:r'əšk'i-ʔí	'roots'	1	=	=	=
(4)	ko:r'əšk'i-ʔí	'roots'	3	=	=	=
(5)	r'imn'í	'belts'	1	r'em'én' r'emn'í	[r'ɪm'én'] [r'ɪmn'í]	m, 2
(6)	r'imn'i-ʔí	'belts'	2	=	=	=
(7)	r'e:mn'i-ʔí	'belts'	1	=	=	=
(8)	o:gurcɨ-ʔí	'cucumbers'	1	ogur'éc ogurcí	[egʊr'éc] [egʊrcí]	m, 2

Recall the forms based on Rus. kor'ešk'í 'roots' discussed above in §5.2. Both singular and plural bear final stress in Russian, thus the vowel in the first syllable is always unstressed and rendered as [a] in Kamas, as in (Table 22: 2) (Strategy C). However, stem-final stress is not retained in forms following Strategy B; the lengthening in these forms is shifted to the first vowel, which is alternatively realized as [aː] (3) or as [oː] (4), the latter corresponding to the underlying vowel of the root, also seen in (1). In a similar vein, the plurals based on Rus. *r'emn'i* 'belts' show alternatively [i] (with no prominence on root) and lengthened [eː] under stress shift. Finally, the only occurrence of *oːgurcɨ-ʔi* 'cucumbers' also features an initial stress with the vowel restored to [oː] in the root where Standard Russian has always [ɐ], and /o/ is only etymological.

To sum up, I argue that the forms in (Table 19: 1–6) are built on Russian Nom. sg. stem with stress shifted to the first syllable and the corresponding vowel restored to its underlying quality, undoing the effects of Russian reduction. I will classify these forms as a particular case of Strategy A, denoting it as 'A-shift'.

5.4.3 Russian masculine nouns with fleeting vowel: A generic oblique stem?

Our third group of interest includes Russian masculine nouns with a consonant-final root. A common feature in this group is the fleeting vowel present only in Nom. sg., creating a root-final consonant cluster in all other forms including Nom. pl. Here as well, a distinctly pronounced low vowel [a] appears stem-finally before the Kamas plural suffix, except in *n'eːms*[ə]-*ʔi* 'Germans'. See examples in Table 23:

Table 23: Russian masculine nouns with fleeting vowel.

No	Kamas		Count	Russian			
	Sing. | Plural	Translation		Nom. sg.	Gen. sg.	Nom. pl.	Gram.
(1)	—			bot'ínok	bot'ínka	bot'ínk'i	m, 2
	bat'iːnka-ʔi	'shoes'	4	[bet'ínək]	[bet'ínke]	[bet'ínk'i]	
(2)	—			estón'ec	estónca	estónci	m, 2
	estoːntsa-ʔi	'Estonians'	1	[ɪstón'ɪc]	[ɪstónce]	[ɪstóncɨ]	
(3)	—			kolód'ec	kolódca	kolódci	m, 2
	kaloːtsa-ʔi	'wells'	1	[kɐlód'ɪc]	[kɐlótce]	[kɐlótcɨ]	
(4)	—			mordóv'ec	mordóvca	mordóvci	m, 2
	mardoːfsa-ʔi	'Mordvins'	1	[merdóv'ɪc]	[merdófce]	[merdófcɨ]	
(5)	n'ém'is	'German'	1	n'ém'ec	n'émca	n'émci	m, 2
	n'eːmsə-ʔi	'Germans'	2	[n'ém'ɪc]	[n'émce]	[n'émcɨ]	
(6)	—			z'úbr' (dial.[15])	z'úbr'a	z'úbr'i	m, 2
	zuːbr'a-ʔi	'deer (pl.)'	1	[z'úbr']	[z'úbr'ɪ]	[z'úbr'i]	

---

15 More exactly, Std. Rus. *iz'úbr'* / dial. Rus. *z'ubr'* denotes here an Altai maral.

The nouns in Table 23 form their plurals differently from those considered before. They will therefore be classified as Strategy D. However it turns out to be problematic to identify the stem on which these plural forms are based. Consider the hypothetical forms based on nominative singular and plural stems:

(10) a. *bat'i:nag-ə?i 'shoes' < bat'i:nak+/ə/+?i < Rus. bot'ínok (Nom. sg.)
      b. *bat'i:nk'i-?i 'shoes' < bat'i:nk'i < Rus. bot'ínk'i (Nom. pl.)

(11) a. *esto:n'iz-[u]?i 'Estonians' < esto:n'is+/ə/+?i < Rus. estón'ec (Nom. sg.)
      b. *esto:nsɨ-?i 'Estonians' < esto:nsɨ < Rus. estónci (Nom. pl.)

Words in this group are clearly not based on the Nom. sg. stem, which can be seen from the absence of the fleeting vowel. However, the final [a] is again different from the Nom. pl. suffix. As for the nouns examined in the previous section, this [a] is arguably different from the epenthetic /ə/, since the preceding stressed /o/ would yield a rounding of /ə/ in 3 of 5 roots. Thus a corresponding derivation from a consonant-final Russian stem (i.e. a root with no inflectional vocalic suffix) would have been (11c):

(11) c. *esto:ns-[u]?i 'Estonians' < esto:ns+/ə/+?i
          < Rus. estónc- (plural or oblique singular root)

Note also that the shape of the first form, bat'i:nka-?i 'shoe-PL', does not demonstrate the obligatory palatalization of the Nom. pl. stem (Rus. bot'ínk'i 'shoes'), which is typically retained in Kamas. Instead, it matches most closely another form in the paradigm, namely Genitive singular (Rus. bot'ínka 'shoe.GEN').[16] The same grammatical form also matches the other nouns in Table 23 (cf. Rus. kolódca 'well. GEN', estónca 'Estonian.GEN'). The stem of n'e:msə-?i may be considered ambiguous between Gen. sg. and Nom. pl.

In addition to Table 23, we find a form of similar phonotactic structure, žalu:t-kə́ 'stomach' (used as singular). The source is a variant of Std. Russian želúdok [žɨ̞lúdək], with a fleeting vowel: Gen. sg. želúdka [žɨ̞lútkə], Nom. pl. želúdk'i [žɨ̞lút-k'i]. Here again, Genitive singular is the most likely candidate.

From the functional point of view, note that for most lexemes the same form as Gen. sg. (also referred to as paucal) appears in Russian in numeral phrases with

---

16 As pointed out by Sergey Kniazev (p.c.), there is also a possibility that this word is feminine in Plotnikova's Russian (Nom. sg. bot'inka, Nom. pl., Gen. sg. bot'ínk'i). It would then belong with the regular plurals of the Strategy A. Unfortunately, there are no corresponding forms in the corpus to check.

'lower' numerals ('2', '3' and '4'), which might have reinforced its motivation to be recruited in a plural context, e.g. *dva bot'ínka* 'two shoes [paucal]'.[17] For animate masculine nouns, Gen. sg. is also syncretic with Acc. sg., cf. *dva estónca* 'two Estonians [paucal]', *ja vstr'ét'ila estónca* 'I met an Estonian [Acc. sg.]'. One might speculate that the genitive-paucal-accusative form was reinterpreted by Plotnikova as a generalized oblique stem and used when the nominative stem was dispreferred (hypothesis proposed by V. Gusev, p.c.).[18]

What is then the motivation to disprefer the nominative stems for the words listed in Table 23? While no straightforward solution is evident, I hypothesize that the answer lies in the prosodic/phonotactic domain. The consonant-final Nom. sg. stem would require an epenthetic /ə/, and the stress on the root would fall relatively far from the accented plural marker (3 syllables away), as can be seen in (10a, 11a). Such forms are in fact encountered in the corpus (e.g. *laːbaz-əʔí* 'storehouses', *koːrʼəšk'i-ʔí* 'roots'), however infrequently (32 tokens, including 16 of *kuːriz-əʔí* 'chickens').

Consider Table 24. The dominating pattern in plurals with suffix -*ʔí* has a distance of two syllables between the Russian stress on the root and the plural suffix, as in *bliːn-əʔí* 'pancake-PL', *naliːm-əʔí* 'burbot-PL', *kaševaːrka-ʔí* 'porridge.cauldron-PL'; over 70% forms in each strategy belong to this pattern.

Table 24: Distance between Russian stress and the plural suffix -*ʔí* in Kamas loans.

Strategy	Distance			
	1 syllable	2 syllables	3 syllables	Total[19]
A	11 (7.0%)	122 (77.2%)	25 (15,8%)	158
B	1 (4.2%)	18 (75,0%)	5 (20,8%)	24
A/B	–	22 (91,7%)	2 (8,3%)	24
A-shift	–	8 (100%)	–	8
Total	**12 (5.6%)**	**170 (79,4%)**	**32 (15,0%)**	**214**

---

17 However, none of the relevant Kamas examples involve a numeral phrase. The lexeme *čas* 'hour' is one of the few lexical exceptions in Russian which take a special form with numerals '2', '3' and '4', *časá*; it is segmentally identical to Gen. sg. (*čása*) but has word-final stress. The only corresponding Kamas example in the corpus has an uninflected Nom. sg. *čas*: *šide čas* 'two hours'.

18 The genitive/oblique stem cannot serve as an explanation for the feminine nouns discussed in §5.4.2, since Gen. sg. is there segmentally identical to Nom. pl. but has word-final stress; cf. Kamas *groːza-ʔí* 'thunderstorm-PL', Rus. Nom. sg. *grozá*, Nom. pl. *grózi*, but Gen. sg. *grozí*, Acc. sg. *grozú*.

19 Table 24 only includes forms with the plural suffix -*ʔí* (and not those with -*zaŋ*), therefore the totals here are lower than the total counts of respective strategies.

Note further that a large share of all the nouns encountered with this stress pattern are feminine nouns of the first declension with a prefinal consonant cluster and final /-a/, e.g. *kni:ška-ʔí* 'book-PL', *ska:ska-ʔí* 'tale-PL'; *p'i:xta-ʔí* 'fir-PL'. The masculine nouns with a fleeting vowel like *bot'ínok* and *želúdok* might have been forced by analogy to select a similar form in their paradigm, avoiding a longer distance between the stress and the accent-bearing plural suffix, and also avoiding more complex phonological adaptations (epenthesis and intervocalic voicing).

5.5 Summary of the plural marking strategies

Plural marking on Russian loans in Kamas as spoken by Klavdiya Plotnikova employs three main strategies (A, B, C) and some less frequent special variants:

Table 25: Plural marking strategies in Russian loans in Kamas.

Strategy	Comment	Tokens	%	Lexemes
A	Rus. Nom. sg. stem + Kamas PL	164	63.6%	83
A-shift	Rus. Nom. sg. with stress shift + Kamas PL	8	3.1%	6
A/B	Rus. Nom. sg. or pl. stem + Kamas PL	25	9.7%	18
B	Rus. Nom. pl. stem + Kamas PL	34	13.2%	16
C	Rus. Nom. pl. stem	16	6.2%	11
D	Rus. generic Oblique stem + Kamas PL	11	4.3%	6
Total		**258**	**100.0%**	**132**

Strategy A is the most frequent one, represented by 164 of the 258 tokens in the corpus, and can be considered the default strategy.

Strategy B is found in another 34 tokens of 16 nouns. Nouns with higher likelihood to occur in plural in the donor language seem to prefer this strategy, such as *pluralia tantum*, nouns denoting pair objects and mass nouns. Also, nouns with word-final Russian stress in singular choose Strategy B if their plural has non-final stress.

Furthermore, 25 tokens of 18 nouns are ambiguous between singular and plural stem in the absence of cues such as final obstruent voicing or stem alternations, or when the cues are contradictory (Strategy A/B).

With 16 tokens of 11 nouns, Strategy C is a minority option. For 8 out of these 11 words, alternative forms exist in the corpus using either Strategy A or B, with an overt Kamas plural marker.

The remaining cases discussed in §5.4.2–5.4.3 demonstrate a complex interaction of the morphological systems of the two languages with some word-prosodic considerations.

The syllable bearing original stress in Russian is typically realized in Kamas with vowel lengthening (and often sustained pitch level). In the presence of Kamas morphology, this can be combined with the accent either on the Kamas suffix itself (e.g. plural -ʔi) or on an adjacent syllable (e.g. before Lative -nə/-də or after the plural -zaŋ). The Kamas accent is principally realized as a tonal contour, while many consonant-initial suffixes trigger lengthening of the preceding vowel. The plural suffix -ʔi does not trigger such lengthening; this is probably the reason why Russian stress in directly preceding syllable is dispreferred.

In feminine nouns with variable stress the preferred solution is not the plural stem which would end in an unstressed [-ɨ] (atypical for Kamas), but the singular stem adapted by shifting the original stress to the left and restoring the corresponding vowel to its non-neutralized quality (Strategy A-shift, 8 tokens of 6 nouns).

For a number of consonant-final masculine nouns with a fleeting vowel, Strategy A would require vowel epenthesis and consonant voicing, resulting in a form with a distance of three syllables between the original stress and the accent on the plural suffix. Such prosodic pattern is infrequent in the corpus. Strategy B would again result in a stem-final unstressed [-ɨ]. The apparent solution (Strategy D, 11 tokens of 6 nouns) employs an oblique stem of genitive singular, also identical to the 'paucal' form used with lower numerals and to the accusative singular of animate nouns. This oblique stem, perhaps supported by the analogy with the very frequent feminine stems, conforms to the two-syllable distance between the original stress and the Kamas accent, at the same time avoiding more complex phonological adaptations.

6 Conclusion

Strategies of plural formation acting in Russian loans in Kamas summarized above cover all the three types identified by Roseano (2014) (see §1.4).

The two extremes are Strategy A (Kamas plural marker only, oikomorphological) and Strategy C (Russian plural marker only, xenomorphological). All the other strategies, i.e. Strategy B, A-shift and D, shall be classified as allomorphological, since they are not limited to the rules of either donor language or target language system. Moreover, instances of Strategy A where the Kamas rules of epenthesis or intervocalic voicing fail to apply might also be regarded as allomorphological.

Further distinctions among these strategies concern the shape of the stem to which the Kamas marker can be attached. Forms based on Russian Nominative plural (Strategy B) are instances of double marking, a phenomenon well known for plural on borrowings (Matras 2009: 174, Myers-Scotton 2002: 92). Some languages can treat a borrowed noun with its original plural marking as singular unless a native plural marking is also applied (cf. recent Russian *baks* 'dollar' < English *bucks*; plural *baksï*). Post-shift Kamas has not gone that far, and Russian plural only seems to appear when semantically motivated, perhaps with the exception of the borrowed *pluralia tantum* nouns like *varo:ta-ʔí* 'gate' which have no Kamas-internal motivation to bear plural marking.

The remaining minority types, Strategy A-shift and D, have a certain similarity. In both cases, the shape of the Russian noun is transformed so as to come closer to the majority phonotactic pattern of the Strategy A, namely a vocalic stem ending in /-a/, with a distance of two syllables between the Russian stress and the Kamas plural suffix -ʔi. The analyses suggested here involve stress shift in feminine nouns (A-shift) and a generic oblique stem in masculine nouns instead of the Nominative singular and plural stems (D). Alternative explanations can surely be sought for, the problem being the limited amount of data and variability issues mentioned in §1.5.

A particularly interesting finding was the relation between the vowel lengthening corresponding to the original stress in Russian loans and the automatic vowel lengthening before a consonant-initial Kamas suffix. They are similar in their durational effect; both do not occur before the plural suffix -ʔi. Yet, the former has access to the underlying vowel quality prior to Russian reduction, while the latter only alters the duration of the vowel without changing its quality.

Two issues deserve particular attention for future research. First and foremost, the word prosody of Kamas and its interaction with Russian (as well as word prosody of Plotnikova's Russian *per se*) should be studied in much greater detail to allow for better grounded conclusions, and perhaps to tell more about the evolution of Kamas since Kai Donner's visit. Another interesting question is the conspicuous use of stem-final /-a/, visibly distinct from an epenthetic schwa, scattered across borrowings of different morphological type and different target forms.

Abbreviations

Language attribution of forms

Rus.	Russian
Std. Rus.	Standard Russian
dial.	(Russian) dialectal form

Grammatical categories

Names of grammatical forms of nouns are capitalized, e.g. 'Nom. pl.' for Nominative plural. When a particular morpheme is referred to, e.g. in glosses, its label is written in small capitals, e.g. Russian *kápl'-i* 'drop-NOM.PL', Kamas *ka:pl'a-ʔí* 'drop-PL'.

f	feminine
m	masculine
n	neutral
adj	adjectival declension
sg.	singular
du.	dual
pl.	plural
pl. t.	pluralia tantum (no singular form)
Abl.	Ablative
Acc.	Accusative
Gen.	Genitive
Dat.	Dative
Instr.	Instrumental
Lat.	Lative
Loc.	Locative
Nom.	Nominative
Prep.	Prepositional
Poss.	possessive

Abbreviations in tables

In columns labelled 'Singular | Plural', the upper line for each noun lists the singular form (if present) and the lower line lists the plural form.

Ct.	count (of Kamas forms in the corpus)
Gram.	grammatical features: gender (m, f, n), declension type (1, 2, 3; adj)
Ep.	vowel epenthesis
Fl. v.	fleeting vowel
Pal.	stem-final palatalization
Rd.	schwa rounding
Vc.	stem-final voicing
n/a	process not applicable (not relevant)
—	singular form not encountered in the corpus (Kamas) / does not exist (Russian pl. t.)

References

Arkhipov, Alexandre V., Chris Lasse Däbritz. 2018. Hamburg corpora for indigenous Northern Eurasian languages. *Tomsk Journal of Linguistics and Anthropology* 3(21). 9–18.

Arkhipov, Alexandre V., Chris Lasse Däbritz & Valentin Gusev. 2020. *User's Guide to INEL Kamas Corpus.* (Working Papers in Corpus Linguistics and Digital Technologies: Analyses and Methodology 3). Szeged & Hamburg: University of Szeged, Universität Hamburg.

Avanesov, Ruben I. 1956. *Fonetika sovremennogo russkogo jazyka.* [Phonetics of Modern Russian]. Moscow: Izdatel'stvo Moskovskogo universiteta.

Blinova, Olga I. 1971. O termine "starožil'českij govor Sibiri". [On the term "old dwellers' Siberian dialect"] *Voprosy jazykoznanija i sibirskoj dialektologii* 2 [Issues in linguistics and Siberian dialectology 2]. 3–8. Tomsk.

Bondarko, Liya V. 1977. *Zvukovoj stroj sovremennogo russkogo jazyka.* [The sound system of Modern Russian]. Moscow: Prosveščenie.

Bondarko, Liya V. 1998. *Fonetika sovremennogo russkogo jazyka.* [Phonetics of Modern Russian]. St. Petersburg: St. Petersburg University.

Castrén, Matthias A. 1847. Manuscripta Castréniana XIX. Samoiedica 13: Kamass-Samoiedica. Unpublished manuscript, National library, Helsinki.

Comrie, Bernard, Gerald Stone & Maria Polinsky. 1996. *The Russian language in the twentieth century.* Oxford: Clarendon Press.

Timberlake, Alan. 2004. *A reference grammar of Russian.* Cambridge: Cambridge University Press.

Gusev, Valentin. 2020. O kamasinskom postpozitivnom uslovno-vremennom sojuze *dak.* [On the postpositive conditional-temporal conjunction *dak* in Kamas]. *Proceedings of the V. V. Vinogradov Russian Language Institute* 26(4). 76–88.

Gusev, Valentin, Tiina Klooster & Beáta Wagner-Nagy. 2019. *INEL Kamas Corpus.* Version 1.0. Publication date 2019–12–15. http://hdl.handle.net/11022/0000-0007-DA6E-9. Archived in Hamburger Zentrum für Sprachkorpora. In Beáta Wagner-Nagy, Alexandre Arkhipov, Anne Ferger, Daniel Jettka & Timm Lehmberg (eds.), The INEL corpora of indigenous Northern Eurasian languages. https://hdl.handle.net/11022/0000-0007-F45A-1.

Iosad, Pavel. 2012. Vowel reduction in Russian: No phonetics in phonology. *Journal of Linguistics* 48(3). 521–571. DOI: 10.1017/S0022226712000102.

Joki, Aulis J. 1944. *Kai Donners Kamassisches Wörterbuch nebst Sprachproben und Hauptzügen der Grammatik.* (Lexica Societatis Fenno-Ugricae VIII). Helsinki: Suomalais-Ugrilainen Seura.

Jones, Daniel & Dennis Ward. 1969. *The phonetics of Russian.* Cambridge: Cambridge University Press.

Klumpp, Gerson. 2022a. Kamas. In Marianne Bakró-Nagy, Johanna Laakso & Elena Skribnik (eds.), *The Oxford Guide to the Uralic Languages*, 817–843. Oxford: Oxford University Press.

Klumpp, Gerson. 2022b. Assessing Heritage Kamas. [Unpublished materials for the talk presented at *Congressus Internationalis Fenno-Ugristarum XIII*, Vienna, 21–27.08.2022].

Knyazev, Sergey V. 2006. *Struktura fonetičeskogo slova v russkom yazyke: sinxronija i diaxronija* [The structure of phonetic word in Russian: synchrony and diachrony]. Moscow: MAKS-press.

Knyazev, Sergey V. & Sofya K. Pozharickaja. 2011. *Sovremennyj russkij literaturnyj jazyk: fonetika, grafika, orfografiia, orfoepija.* [Modern Literary Russian: Phonetics, Graphics, Orthography, Orthoepy]. 2nd ed. Moscow: Akademicheskij proekt; Gaudeamus.

Matras, Yaron. 2009. *Language Contact.* Cambridge: Cambridge University Press.

Matveyev, Aleksandr K. 1965. Novye dannye o kamasinskom jazyke I kamasinskoj toponimike [New data on Kamas language and Kamas toponymy]. *Voprosy toponomastiki* 2 [Issues in topono-mastics 2]. Sverdlovsk: Izdatel'stvo Ural'skogo universiteta. 32–37.

Myers-Scotton, Carol. 2002. *Contact Linguistics*. Oxford: Oxford University Press.

Roseano, Paolo. 2014. Can morphological borrowing be an effect of codeswitching? Evidence from the inflectional morphology of borrowed nouns in Friulian. *Probus* 26(1). 1–57. DOI: 10.1515/probus-2013-0007.

SRNG 2012 = *Slovar' russkix narodnyx govorov.* [Dictionary of Russian folk dialects]. Vol. 45. St. Petersburg: Nauka, 2012.

Yanushevskaya, Irena & Daniel Bunčić. 2015. Russian. *Journal of the International Phonetic Association* 45(2). 221–228. DOI:10.1017/S0025100314000395.

Zaxarova, Kapitolina F. & Varvara G. Orlova. 2004. *Dialektnoe členenie russkogo jazyka.* [Dialectal partitioning of Russian language]. Moscow: Editorial URSS.

Zlatoustova, Lyubov V. 1956. Fonetičeskaja priroda russkogo slovesnogo udarenija [Phonetic nature of Russian word stress]. *Uchenye zapiski Kazanskogo gos. unversiteta* 116(11). 3–38.

Zaliznyak, Andrey A. 1967. *Russkoe imennoe slovoizmenenie.* [Russian nominal inflection]. Moscow: Nauka.

Thomas Stolz

4 On co-plurals

Cross-linguistic evidence of competing pluralization strategies
in the domain of nouns

Abstract: The paper investigates the recurrence of competing plurals across languages of different genealogical affiliation as well as different typological and geographic backgrounds. It is shown that the phenomenon is widespread. In some languages, it affects numerous nouns. Competing plurals may be used to differentiate meanings of otherwise polysemous nouns. There is also ample evidence of the absence of genuinely semantic distinctions. Style and register are often decisive factors for the choice of plural.

Keywords: overabundance, competition, style, semantics, plural

> To the memory of the late Wilfried Fiedler
> (*1933–†2019), eminent Albanologist

1 Introduction

In a still undetermined number of languages at least some nouns allow for the coexistence of two or occasionally more than two plurals. Henceforth, I use the term COPL[urals] to refer to the competing expressions. The phenomenon as such is not a brand-new discovery. Students of Ancient Greek, for instance, have always been familiar with doublets such as SG *génos* 'lineage' → PL *génea* ~ *génē* (Bornemann and Risch 1978: 44) which reflect the differences between Attic and other varieties of Ancient Greek. However, by no means all instances of one-to-many relations of

---

Acknowledgments: This paper elaborates on the ideas expressed in the talk entitled *On the coexistence of multiple pluralization strategies – an explorative cross-linguistic survey* which I delivered on occasion of the pre-conference workshop on *Number Categories* (3 June, 2021 in Bremen/Germany). I am grateful to my audience – especially Isabel Compes, Jürg Fleischer, and Paul Widmer – for their thought-provoking comments. Some of the ideas expressed in this paper have arisen from discussion with Benjamin Saade. Anna Thornton provided me with invaluable reading matter and commented on the draft version of this paper. Her comments induced me to revise some of my original hypotheses. Greville Corbett kindly commented on the draft version of the paper. Thanks to Deborah Arbes' detailed criticism of the original version, I was able to produce a much better final version of this paper. Nataliya Levkovych and Maike Vorholt kindly gave me technical support. All errors and mannerisms which remain are, however, exclusively mine.

https://doi.org/10.1515/9783110986600-004

singulars and plurals can be explained in terms of diatopic variation. Within the framework of Canonical Morphology, Thornton (2010) – taking up an issue raised by Acquaviva (2008: 123–161) – describes COPLs of the type SG *braccio* 'arm' → PL *braccia* ~ *bracci* in Italian as instances of noncanonicity which mostly correspond to overabundance (i.e. a given noun is equipped with two synonymous plurals) and to a minor extent, depending on the linguist's approach, also defectivity (i.e. one of the semantically distinct plurals lacks a corresponding singular). The third possibility according to which COPLs realize the phenomenon of overdifferentiation (i.e. the COPLs reflect a distinction of additional number values) is ruled out for Italian. The term COPL deliberately glosses over the differences between these categories of noncanonicity in order to allow for an unprejudiced discussion of the phenomena to be encountered.

The notion of COPL covers a wide range of situations. Two expressions X and Y are COPLs of each other if they realize the feature PLURALITY for the same noun. COPLs can be pervasive in a given system in the sense that all nouns and all appropriate cells of the paradigm are affected by the phenomenon. Most often one of the COPLs can be characterized as associative plural, distributive plural, or collective (Corbett 2000: 101–120). However, in this paper, the focus is on those situations in which COPLs constitute mismatches because their domain covers only selected nouns and/or cells of paradigms.

In the conclusions of her seminal paper on noncanonical plurals in Italian, Thornton (2010: 56) argues that

> [c]ome avviene per tutti i fenomeni non canonici, non ci si aspetta di trovare un numero alto di esempi in ogni singola lingua, né nelle lingue del mondo nel loro insieme, ma lo studio dei pochi esempi individuabili può contribuire a comprendere quali forme possa assumere il paradigma di un elemento lessicale.[1]

This quote depicts COPLs as a potentially marginal phenomenon. However, since a comprehensive and typologically inspired account of multiple options for the pluralization of nouns is still wanting,[2] we can neither be sure that COPLs are indeed

---

1 My translation: "As happens with all noncanonical phenomena, one does not expect to find numerous examples in every single language nor in the entirety of the world's languages, but the study of the few identifiable examples may contribute to understand what shape the paradigm of a lexeme might take."

2 The bulk of the literature on overabundance focuses on verb morphology or touches on COPLs only in passing (Thornton 2010, 2012a–b, 2013, 2019a–b). As to lexical plurals (which constitute a subclass of COPLs), the situation is different since Acquaviva (2008) provides case studies on related phenomena in Italian, Irish, Breton, and Arabic. He also touches unsystematically upon a number of pieces of evidence for COPLs in a wide variety of languages without however exhausting the topic.

infrequent cross-linguistically nor that they generally behave the Italian way. Moreover, we are still largely in the dark about their raison d'être. Acquaviva (2008: 33) claims that

> grammatically equivalent plural alternants are exceptional but not impossible, and when they arise languages tend to semantically differentiate them. When a noun has more than one plural form, therefore, the alternants are no longer the grammar-driven outcome of an inflectional process, but involve a lexical choice.

The quoted author distinguishes stylistic, contextual, semantic, and semantic-grammatical motiviations for COPLS deliberately discounting diatopic variation, sociolinguistic variation, and archaisms because these factors do not contribute to the genesis of lexical plurals (Acquaviva 2008: 34). I strongly argue that to adequately capture the nature of COPLS we have to avoid focussing exclusively on those cases which involve semantic differentiation (= lexical plurals in Acquaviva's approach). With reference to data from Hausa and Dholuo, Storch and Dimmendaal (2014: 3–6) emphasize that COPLS need to be investigated in-depth because the authors assume that there is a linguistically very intriguing network of factors which motivate the coexistence of several plurals – a network that calls for being disentangled. This study is intended as a first step towards a full-blown cross-linguistic inquiry into the nature of COPLS.

Before I explain why an investigation of this kind makes sense linguistically in the first place, I illustrate the phenomenon with two sentential examples from the Afro-Asiatic language Maltese in (1).[3] The examples are taken from the electronic Korpus Malti 3.0 and are representative of two different genres, viz. the language of law (= (1a)) and that of Eurolectal Maltese (= (1b)), i.e. the style which has developed in the context of the EU administration (Portelli and Caruana 2018).

(1) Maltese [mlrs.research.um.edu.mt]

   (a) [Law815]

       fil-vettura   *jkun*   *hemm*   *imħolli*   *apposta*   *wisa'*

       in:DEF-vehicle   3:be.FUT   there   PASS.PART:leave on_purpose space

       għal   *dawn*   *il-pakki*   *jew*   ***sarar***

       for   DEM:PL:PRX   DEF-parcel:PL or   **bundle.PL**

       '. . .in the vehicle there will purposely be left space for these parcels or **bundles. . .'**

---

3 In the numbered examples, the competing plurals are marked out in boldface. The morpheme glosses follow the guidelines of the *Leipzig Glossing Rules*.

(b) [european8781]

L-Istati	*Membri*	*għandhom*	*jeħtieġu*	*illi*	*l-pakketti*
DEF-state:PL	member:PL	have:3PL	3:need:PL	that	DEF-parcel:PL
u	*s-soror*		*tal-materjal*	*tal-propagazzjoni*	*jkunu*
and	DEF-**bundle.PL**		of:DEF-material	of:DEF-propagation	3:be.FUT:PL
ssiġillati	*uffiċjalment*				
seal:PASS.PART:PL	officially				

'. . .the member states need to take care that the parcels and **bundles** of the propagation material are officially sealed. . .'

The plural of the noun *sorra* 'bundle' is realized as *sarar* in (1a) and *soror* in (1b). The coexistence of *sarar* and *soror* is not determined by any phonological rule. *Sarar* and *soror* form a pair of COPLs. These wordforms reflect the patterns XIII 1*a*2*a*3 and XVI 1*o*2*o*3 of the broken plural in which the sequence of the three root consonants 1-2-3 is broken up in two syllables by way of intercalating the vowels *a-a* and *o-o*, respectively (Borg and Azzopardi-Alexander 1997: 179–180). There are other nouns which pluralize regularly according to pattern XIII (e.g. SG *baqra* 'cow' → PL *baqar*) or pattern XVI (e.g. SG *triq* 'street' → PL *toroq*).

The COPLs *sarar* and *soror* occur as second conjunct of a disjunctive (= (1a)) or conjunctive (= (1b)) coordination in combination with the (sound, i.e. suffixal) plural of semantically related nouns, namely *pakk* 'parcel' and *pakkett* '(small) parcel' both pluralized by way of adding the suffix -*i*. I argue that *sarar* and *soror* could replace each other in the two sentences because there is no discernible semantic difference between them. There is no reason either to assume that the different genres (law vs EU administration) or the different types of coordination (disjunctive vs conjunctive) or the presence/absence of the definite pro-clitic on the pluralized noun has a say in the choice of the plural form. The two options of pluralizing *sorra* must be familiar to most if not all native speakers of Maltese at least passively. Whether one and the same speaker makes use actively of both plurals cannot be determined in this paper. According to Aquilina (1991: 1275) both broken plurals are attested already in the 18th century.

In the Korpus Malti 3.0, the singular *sorra* occurs 60 times whereas the token frequency is down to 13 hits for *sarar* and 9 hits for *soror*. The absolute numbers are too small to draw far-reaching conclusions on their basis. What can be said nevertheless is that the two plurals seem to be fully synonymous of each other with both plurals being almost identically infrequent. We thus have a number paradigm with two cells – one each for the singular and plural of *sorra*. The cell reserved for the plural hosts two occupants which have the same meaning and which, as far as I can tell on the basis of the electronic resource, also largely share the same distribution (cf. below). In this case, it is legitimate to speak of overabundance involving two cell-mates which constitute a case of free variation (Thornton 2019a: 223).[4] As we will discover in the course of this study, free variation is not the default case for COPLS.

There is a very slim chance that even the initial Maltese examples could be interpreted differently. The second meaning of *sorra* is 'flank (of a tunny)'. In Schembri's (2012: 84 and 112) study of the broken plural in Maltese, *sorra* is exclusively registered with this meaning and the plural is always given as *soror* (the alternative *sarar* is not mentioned at all). In the Korpus Malti 3.0, the second meaning is not attested. To determine whether the different plurals tend to associate with different meanings of the noun, an in-depth study has to be carried out on the basis of a much larger corpus. For the sake of the argument, I assume that *sarar* is ruled out as plural of *sorra* if the intended meaning is 'flanks.' What we learn from the still very superficial discussion of the above Maltese data is that straightforwardness cannot be expected to apply to all instances of COPLS.

The Maltese case raises the question whether the coexistence of alternative plurals of one and the same noun is incidental. Superficially, having two expressions for the same meaning seems to violate the principles of economy (Vulanović and Ruff 2018) in the sense that one plural suffices to fulfil the task of number marking. A second plural therefore appears to be disposable because it is at odds with the principle of one-meaning-one-form (Miestamo 2008: 32–33). Thornton (2019b: 13–17) critically reviews a plethora of further principles which seem to bar the possibility of overabundance to emerge and persist (especially in first language acquisition[5]). At the same time, one might interpret the presence of several plurals to be indicative of functional differences (= overdifferentiation) between the options according to a line of thinking which Givón (1985: 10) denounces as "naïve functionalism" because it denies the possibility of any kind of arbitrariness to apply in form-function pairings. The competition of several plurals thus poses a variety of problems whose solution is relevant to linguistic theory as shown in Thornton (2019a). The primary aim of this paper is to instigate dedicated follow-up studies by way of demonstrating that COPLS recur cross-linguistically far too often

---

4 I take it for granted that Thornton tacitly presupposes that only those utterances should be taken account of which are produced by fully competent native speakers of the object language (in her case, native speakers of Italian). I share this implicit point of view whereas Storch and Dimmendaal (2014: 4) also mention the learner varieties of speakers who have "not yet gained firm competence in Dholuo grammar." The (nonstandard) overuse of the pluralizer *-ní* by learners of Dholuo creates additional COPLS as in SG *díèl* 'goat' → PL *díégì* ~ *dìè:k* (~ *díèlní*[LEARNER_VARIETY]). Note additionally that different sources register different plurals for Dholuo which might reflect dialectal differences.
5 How children master COPLS in first language acquisition is also a major concern of Storch and Dimmendaal's (2014: 4–6).

to be filed away as marginal curios. COPLs are attested in languages of different genealogical, typological, and areal background. This diversity notwithstanding, there are also certain properties associated with COPLs across several unrelated or only distantly related languages which suggest that the phenomenon is not devoid of systematicity albeit a restricted one since Thornton's (2010) Italian COPLs suggest that many lexemes behave individually so that generalizing over them becomes problematic.

The paper is organized as follows. In Section 2, the story of COPLs in Maltese is continued to provide us with the essential guidelines for the further elaboration of the subject at hand. Section 3 focuses on a selection of cases of supposed free variation of competing plurals. Section 4 is reserved for those cases of COPLs which involve semantic differentiation – either grammatical or lexical. The case-studies presented in the empirical part of the paper only scratch the surface of the phenomenon. The (preliminary) conclusions are drawn in Section 5. As to theory and methodology, I take sides with the canonical approach in typology as propagated by Corbett (2005). For the taxonomy of number categories, I rely on Corbett (2000). In all matters related to overabundance, overdifferentiation, and defectivity, I am indebted to the ideas expressed in Thornton's (2010, 2012a–b, 2013, 2019a–b) oeuvre to which I will refer time and again in the subsequent sections.[6] No formal sampling procedure has been applied to the languages from which I draw the necessary examples. Only languages which give evidence of COPLs are taken account of. The data stem for the most from the usual descriptive linguistic literature to which corpus data are added occasionally (mostly sentential examples). The methodology is overwhelmingly qualitative adorned with some unsophisticated frequency counts. Similarly, instances of COPLs are approached from the synchronic point of view without excluding diachronic digressions completely.

2 COPLs in Maltese

Maltese nouns are formally sensitive only to number and related categories (singular, plural, counting plural (paucal), collective, singulative, dual). In their reference grammar of Maltese, Borg and Azzopardi-Alexander (1997: 136) mention that "some nouns have the two types of plural," meaning: a sound plural and

---

6 Since Thornton (2010) already conducted an in-depth study of COPLs in Italian, I refrain to from reinventing the wheel in this study, meaning: Italian does not feature among the case-studies included in this paper.

a broken plural coexist as options for one and the same noun as e.g. SG *tapit* 'carpet' → PL *tapiti ~ twapet*. The COPLs *sarar ~ soror* are by no means an isolated case in the grammatical system of Maltese. The Maltese grammarians further argue that

> [s]ome singular forms take two different plurals []. Such plurals may be in free distribution within a particular idiolect, or else one and not the other, may occur in a given idiolect [.] Sometimes two different plural forms serve to disambiguate a homonymous singular form (Borg and Azzopardi-Alexander 1997: 186).

In Table 1, I reproduce the typical patterns of COPLs according to the description provided by Borg and Azzopardi-Alexander (1997: 174–188). Grey shading singles out those cells which host a broken plural. All other plurals are sound or, in the case of *trufijiet* 'edges', a combination of broken and sound plural.

Table 1: Selected COPLs in Maltese.

singular	meaning	plural 1	plural 2
werqa	'leaf'	*werq-at*	*werq-iet*
giddieb	'liar'	*giddieb-a*	*giddib-in*
arġentier	'silversmith'	*arġentier-a*	*arġentier-i*
ċekk	'cheque'	*ċekk-s*	*ċekk-ijiet*
flett	'flat'	*flett-s*	*flett-sijiet*
termos	'(hot water) flask'	*triemes*	*termos-ijiet*
tarf	'edge'	*truf*	*truf-ijiet*
kaxxa	'box'	*kaxex*	*kaxx-i*
kutra	'blanket'	*kwieter*	*kutr-i*
riħ	'wind; a cold'	*rjieħ* 'winds'	*riħ-at* 'colds'
bank	'bank; bench'	*banek* 'banks'	*bank-ijiet* 'benches'

Table 1 gives evidence of the competition of different plural suffixes and of broken plurals which compete with sound plurals. Only in the latter combination do we find examples of semantic differentiation in the sense that the meanings of a polysemous noun or two homophonous nouns are formally distinguished in the plural. For the overwhelming majority of the cases, Borg and Azzopardi-Alexander's (1997: 183) judgment seems to hold according to which "[t]here does not seem to be any distinction in meaning between the two plurals."

Table 1 calls for a comment. First of all, it is clear that COPLs are not absolutely exceptional in Maltese. Secondly, not only inherited nouns of Semitic stock (e.g. *tarf* 'edge') are affected by COPLs but also Romance (e.g. *kaxxa* 'box') and (generally more recent) English loanwords (e.g. *flett* 'flat'). Thirdly, none of the COPLs in Table 1 can be interpreted as a collective or counting plural (paucal) whereas these minor categories are firmly established in a limited segment of the Maltese lexicon (Mifsud 1996). Furthermore, the list is not closed; the inventory of COPLs in Maltese still has to be carried out. Moreover, Borg and Azzopardi's (1997) account skips the possibility of pairs of COPLS – like the above *soror* ~ *sarar* – which consist only of broken plurals. What is more, no mention is made of cases where more than two plurals compete with each other. Two cases in point are SG *qana* 'irrigation canal' → PL *qonja* ~ *qwieni* ~ *qnajja* (Aquilina 1991: 1119) with three options in the plural and SG *faħal* 'stallion' → PL *ifħla* ~ *fħal* ~ *fħul* ~ *fħula* (Aquilina 1987: 297) with four options.

Thornton (2019a: 240–245) also uses quantitative evidence to determine whether cases of overabundance are canonical in the sense that all cell mates display relatively similar token frequencies. The smaller the gap between the frequencies the better the chances that the COPLS are indeed of a kind. In contrast, a huge quantitative gap is potentially indicative of (semantically, pragmatically, stylistically) conditioned overabundance. Thornton's (2010: 7–8) corpus-based account of Italian COPLS features several cases with very low token frequencies. Moreover, her discussion of selected pairs of COPLS shows that the nouns affected by the phenomenon behave individually so that it is hard to generalize over the entire set of COPLS (Thornton 2010: 39–55). Low token frequency and heterogeneity are leitmotifs in the domain of COPLS also cross-linguistically as the subsequent paragraphs will amply prove.

In the introduction, I have shown that *soror* and *sarar* yield similarly small turnouts so that the quantitative criterion for registering the pair *sarar-soror* as a case of canonical overabundance is met. Is the same result to be expected also for the other COPLS mentioned in Table 1 and the foregoing paragraph? The answer is negative insofar as several factors interfere with the frequency count to yield a very heterogeneous picture. The noun *ċekk* 'cheque' and its plurals are absent from the Korpus Malti 3.0. Similarly, the singular *qana* 'irrigation canal' is attested 10 times but none of the plurals is attested. The turnouts for the plurals *giddieba* 'liars' (91 matches) and *arġentiera* 'silversmiths' (34 matches) cannot be taken as is because there is syncretism with feminine singular nouns and adjectives. The plurals *triemes* '(hot water) flasks', *kwieter* 'blankets' and *fħul* 'stallions' could not be found in the Korpus Malti 3.0. Moreover, a third COPL of *flett* 'flat' showed up, namely *flettijiet* which, with 52 matches, is the most common option in comparison to *fletts* with 14 matches and the hapax *flettsijiet*. Figure 1 captures the shares of the different COPLS of the seven Maltese nouns for which reliable calculations could be conducted.

The only (relatively) balanced relation between two COPLS of a given noun is that of (SG *faħal* 'stallion' →) PL *ifħla* (10 matches) and PL *fħula* (8 matches). In all other cases, there is a quantitatively predominant option which is responsible for the vast majority of all attested plurals of a given noun. The predominance ranges

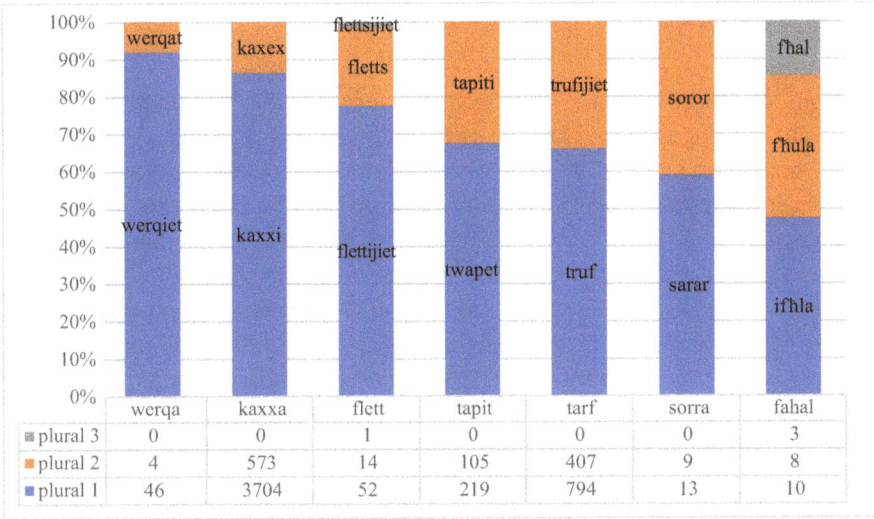

Figure 1: Frequencies of COPLS of seven Maltese nouns.

from 59% in the case of (SG *sorra* 'bundle' →) PL *sarar* to 92% in the case of (SG *werqa* 'leaf' →) PL *werqiet*. It is worth noting that the majority options involve both broken plurals and sound plurals. Third options (where they exist) only claim small shares. There must be factors which determine which of the COPLS is opted for in a given context. The inquiry into this issue is beyond the purpose of this paper since it requires a much bigger database and must involve by far more COPLS. I therefore relegate this pending task to a dedicated follow-up study.

The Maltese data clearly show that the adequate description and evaluation of COPLS cannot be achieved on the basis of what the extant descriptive grammars tell us. There is a heavy dose of heterogeneity and variation on several parameters which can be controlled only by way of conducting a corpus-based investigation. We also learn from the above facts that the pluralization of loanwords is an important factor contributing to the emergence of COPLS. Table 1 hosts COPLS in which the English pluralizer *-s* or the Italian pluralizer *-i* are employed whereas Semitic suffixes like *-ijiet* and *-at* or introflexion are used elsewhere. Matras (2009: 212–213) states that the borrowing of inflectional morphology is generally rare in comparison to that of derivational morphology with the notable exception of the expression of plurality. In Sections 3–4, I have occasion repeatedly to mention COPLS which have arisen in language-contact situations.[7]

---

7 For a different domain (viz. that of ordinal numerals), Stolz (2001) and Stolz and Robbers (2016) demonstrate that overabundance is frequently the by-product of borrowing.

3 "Meaningless" COPLS

The point of departure of the cross-linguistic survey is the "meaningless" kind of COPLS. The reading I suggest for the attribute "meaningless" is that the coexisting plurals neither associate with different lexical meanings nor with different grammatical categories. The opposite possibilities are looked into in Section 4. A pair of "meaningless" COPLS is not necessarily devoid of differences on the connotative, stylistic, or diatopic level. For several of the Italian COPLS, Thornton (2010) shows that there are sometimes subtle preferences for one of the competing plurals over the other in certain contexts albeit without barring completely the other plural(s) from occurring in the very same context. The nominative plural of Polish human (= masculine) nouns ends in -owie (e.g. SG *król* 'king' → PL *królowie*), -*i*/-*y* (e.g. SG *olbrzym* 'giant' → PL *olbrzymi* / SG *aktor* 'actor' → PL *aktorzy*), or -*e* (e.g. SG *marynarz* 'sailor' → PL *marynarze*). For many (but not all) nouns with stems in -*on*, -*un*, -*er*, -*or*, and -*og*, however, plurals in -*owie* and -*i*/-*y* are equally licit so that COPLS are common such as SG *profesor* 'professor' → PL *profesorowie* (n >1,000 hits[8]) ~ *profesorzy* (185 hits) / SG *opiekun* 'legal guardian' → PL *opiekunowie* (n>1,000 hits) ~ *opiekuni* (23 hits). Semantically, the COPLS are identical but "[d]ie Formen auf -**owie** [original boldface] haben für gewöhnlich einen mehr feierlichen Charakter" ['usually the forms in -*owie* have a more ceremonial character'] (Laskowski 1972: 42). Their semantic equivalence notwithstanding, the different forms of the plural are associated with different registers and styles. On the level of connotations, the choice of plural is thus meaningful to the members of the Polish speech community. To identify preferences of this kind, extended corpus-studies are called for – for obvious reasons, this is a task this paper cannot fulfil. At the same time, I try to avoid reading too much into the raw data which are presented in the literature because the hypotheses based thereon would probably all be fallacious.

A typical situation is the following. In the reference grammar of Hausa, the reader's appetite for the topic of pluralization of nouns is whetted by way of presenting the ten COPLS of the noun SG *kádòo* ~ *kádàa* 'crocodile' which I reproduce in (2).[9] Note that there is overabundance in the singular too.

(2)   COPLS of Hausa SG *kádòo* ~ *kádàa* 'crocodile'
  (a) *kádándúnàa*, (b) *kádáadúnàa*, (c) *káddúnàa*, (d) *kádáadúwàa*, (e) *kádóodíi*,
  (f) *kádàndáníi*, (g) *kàdàníi*, (h) *kàddái*, (i) *kàdànnúu*, (j) *kàdùnníi*

---

8 The token frequencies result from a search on the National Corpus of Polish executed on 13 December, 2021 (The full NKJP corpus (1800 M segments)).
9 High-tone diacritics added according to Storch and Dimmendaal (2014: 4).

Wolff (1993: 143) uses this assortment of COPLS to support his idea that in modern Hausa[10] the formation of plurals is no longer predictable on the basis of the singular and that phonological criteria alone are insufficient as determining factors.[11] Storch and Dimmendaal (2014: 4) take this striking case to argue that

> [n]ot every speaker of Hausa will include all of these ten forms in his or her repertoire, as some of them may just be regional variants. However, there is clearly a certain possibility of making choices, and each plural form that a speaker is able to select could have a specific connotation, either in terms of how specific a speaker wants to be by choosing a more marked form, or in terms of register, lect, and so on.

What we learn from this quote is that the distribution of the COPLS in (2) over the Hausa speech-community and its diatopic, diastratic, and diaphasic properties are largely unclear.[12] Similarly, what semantic or connotational differences might exist between the different COPLS cannot be determined in a satisfactory way. Moreover, it is doubtful that Hausa speakers generally know all ten COPLS some of which are probably characteristic of a given variety and some varieties might even avoid COPLS by way of monopolizing the use of only one of the options. On this basis, it is impossible to decide whether and to what extent we are facing cases of overabundance and/or overdifferentiation. It cannot be denied that the many forms in (2) have a certain potential for functional differentiation as alluded to by the quoted authors. However, tangible proof of the "meaningful" uses to which the alternative plurals are put is not available. This is why one is running the risk of getting lost in speculation if uncommented lists of data like (2) are interpreted as hard evidence of the functional specialization of their members.

In Pashto (Indo-Iranian), there are different inflection classes. The formation of the plural is variegated. There are animacy-based semantic criteria which inter-

---

10 For another Chadic language, Kwami, Leger (1994: 127) registers the coexistence of competing plurals which do not reflect meaning differences (some of the nouns with COPLS are borrowings from Hausa). A case in point is SG *tárá* 'goat' → PL *térní* ~ *térníshíná*. Leger (1994: 122) argues that COPLS may arise because of the competition of several pluralization strategies in Kwami and adds that the choice of plural largely depends on the speaker's idiolect.

11 It remains unclear which of the ten wordforms are purely phonological variants of each other and can be explained as the result of the application of general rules which create short forms out of long forms or have regular effects on the tonal structure of the wordforms. For the competition of long and short forms of personal pronouns in Maltese for instance, Stolz and Saade (2016) and Saade (2018) argue that it is impossible to postulate rules which exhaustively capture the distribution of the competing wordforms.

12 This is different in Burmese for instance, because the co-existence of the pluralizers *-dwe* and *-myà* is clearly a matter of register. According to Jenny and Hnin Tun (2016: 128), *-myà* is used exclusively in literary Burmese (although it remains unclear whether its competitor is banned from the written register).

act with segmental and suprasegmental factors when it comes to determining the plural of a given noun (Lorenz 1982: 40–42). Nouns may attest to COPLs. Nonhuman animates may take either the pluralizer -*ə́n* for animates or the corresponding pluralizer -*úna* for inanimates as e.g. SG *ās* 'horse' → PL *āsúna* ~ *āsān*. Animate nouns ending in -*aj* have the choice between the pluralizers -*i* and -*ijān* as e.g. SG *māháj* 'fish' → PL *māhí* ~ *māhijə́n*. Monosyllabic nouns hosting the vowel /a/ add the plural suffix -*ə* or -*una* as SG *war* 'door' → PL *wrə* ~ *wruna*. Those nouns which end in either -*ā* or -*o* have the choice between -*wi* and -*gə́ni* as plural markers as e.g. SG *ārzó* 'wish' → PL *ārzówi* ~ *ārzogə́ni*. The source I consulted does not mention semantic differences between the COPLs. Occasionally, the author remarks that one of the options is more frequent than the other. On the basis of the little information one can get from the description of the plural formation in Pashto, it is impossible to determine whether we are dealing with balanced or conditioned overabundance. This Pashto case is not unique but rather symptomatic of the general situation in the sense that the descriptive linguistic literature usually does not provide enough clues to discover what hides behind the COPLs functionally.[13]

Like Maltese, many languages introduce foreign plural markers to their system in the course of (often massive) lexical borrowing from another language. There are two scenarios which witness the rise of COPLs. Either the autochthonous pluralization strategies of the replica language diffuse into the domain of those loanwords which are borrowed along with their original pluralizers or inversely the borrowed pluralizers spread from the class of loanwords to enter that of the inherited nouns of the replica language. In both cases, some nouns will allow for the co-existence of foreign and autochthonous ways of pluralization at least temporarily. A case in point are Italian loanwords in German which give evidence of COPLs such as Italian SG *pizza* 'pizza' → PL *pizze* > German SG *Pizza* 'pizza' → PL *Pizzen* (2.6 million hits[14]) ~ *Pizzas* (83 million hits) ~ *Pizze* (13 million hits); Italian SG *conto* 'bank account' → PL *conti* > German SG *Konto* 'bank account' → PL *Konten* (104 million hits) ~ *Kontos* (12 million hits) ~ *Konti* (16 million hits). Native speakers of German who are (or pretend to be) intimately familiar with Italian culture would opt at least occasionally for the Italian plurals in -*e* and -*i* even in

---

13 This is the case with Kalmyk SG *yalū* 'goose' → PL *yalūd* ~ *yalūs*. Benzing (1985: 119) claims that the variation between the pluralizers -*d* and -*s* occurs "gelegentlich" ['occasionally']. For a small set of Portuguese nouns ending in -*ão*, the form of the plural has not yet been fixed by normative grammar as e.g. SG *sultão* 'sultan' → PL *sultãos* ~ *sultães* ~ *sultões* although the spoken register seems to prefer the plural in -*ões* (Cunha and Cintra 1989: 183). Whether this remark means that the preferences are different in the written register cannot be determined on the basis of the description. **14** The added approximate token frequencies are based on an informal search on Google carried out on 15 September, 2021.

their German discourse. The majority of speakers prefers the *en*-plural and/or the *s*-plural with the former being typical of inherited nouns of Germanic origin and the latter being reserved for loanwords (especially but not only those ending in a full vowel).

Learned plurals for borrowed nouns (e.g. from Latin or Greek) can be used for reasons of prestige to show one's familiarity with the pillars of culture. Gardani (2012: 86) reports to discuss (to the best of his knowledge) "all cases of borrowing that [] have been reported in the literature so far." These borrowed pluralizers have become (weakly) productive in the replica languages in the sense that they attach also to indigenous stems where they compete with established indigenous plural markers such as borrowed Arabic -*āt* and -*ijāt* in Persian where we find examples of the types SG *dastur* 'order' → PL *dasturhā*~INDIGENOUS~ ~ *dasturāt*~BORROWED~ / SG *sabzi* 'vegetable' → PL *sabzihā*~INDIGENOUS~ ~ *sabzijāt*~BORROWED~ (Yousef 2018: 32). Yousef (2018: 354) mentions also the Arabic broken plural "that came along with the loan words, although the Persian plural is possible for all of them and is even much more common", meaning: there are Arabic loanwords in Persian which attest to COPLs with one of them being a broken plural formed according to the rules of the donor language, whereas its competitor is a Persian plural suffix. The author emphasizes that the use of Arabic plural suffixes on Persian stems is often considered to be bad style. The situation in Persian is comparable to that reported for Ottoman Turkish in the early 20th century. Weil (1917: 86–87) reports the existence of about 50 broken plurals which coexisted with Turkish plural suffixes in the speech of "[m]anche Neuerer" ['certain reformers'] to yield pairs of COPLs such as SG *fakir* 'poor' → PL *fukara* ~ *fakirler*. The employment of the suffix -*ler* had political connotations in the sense that it could be interpreted as expressing the speaker's wish of emancipating Turkish culture from the age-old Arabic dominance. Semantically, however, the COPLs were equivalents of each other. The cross-linguistic catalogue of COPLs in which autochthonous and borrowed pluralizers are involved is so rich that it deserves to be presented separately in a follow-up study. For the time being, it suffices to note down that language contact creates favourable conditions for the emergence of COPLs.

Old and new may co-exist in a given system at a given point in time. This is the case with COPLs in Dinka (Nilotic). Andersen (2014: 248–249) reports on free variation of plurals which is attested for many of the nouns he elicited. The author stresses the fact that the singular is never affected by variation. Examples like SG *rǫ́l* 'throat; voice' → PL *rǫ̀t* ~ *rǫ̀ɔl* are explained as retentions of old (= unproductive) patterns of pluralization alongside new ones. As to meaning, there seems to be no difference between the two options. The archaic character of PL *rǫ̀t*, however, may have certain connotations for speakers of Dinka. It should not go unmentioned that Andersen's (2014: 248 footnote 9) findings do not correspond to those of Ladd, Remijsen and Adong Manyang (2009) who discovered hardly any COPLs. According to Andersen, this can be explained with reference to the dialectal differences between the two varieties of Dinka which served as object languages of the two studies.

Regional variation is indeed a topic that needs to be addressed before we can continue with the empirical case-studies. With reference to the plethora of pluralization patterns in Yiddish, Jacobs (2005: 163) mentions that there is "significant regional variation." If we say that Yiddish SG *noz* 'nose' displays three plurals, namely *nezər*, *nez*, and *nozn*, does this imply that there are COPLs? According to Jacobs (2005: 163), only PL *nezər* is standard whereas PL *nez* and PL *nozn* belong to two different dialects. Situations of this kind are difficult to judge. Given that at least two of the three plurals form part of one and the same speaker's native language competence, it is safe to assume that they are COPLs. If, however, the full range of options is known only to the philologist, doubts are unavoidable as to the COPL-status of the wordforms. Ideally, they should belong to the same variety and not to different varieties of the same diasystem. Deciding this issue is easier said than done because the descriptive literature does not usually indicate whether a given option is available only to speakers of a given variety. Accordingly, some of the COPLs discussed subsequently might result from the lumping together of plurals which never co-occur in the same variety of a given language.

The above aspects have to be kept in mind when we look at the sketches of COPLs in Estonian (Section 3.1), Georgian (Section 3.2), Nahuatl (Section 3.3), Breton (Section 3.4), and Dutch (Section 3.5). The focus in these sections is on those cases of COPLs which reflect no meaning distinctions. This does not mean that the languages under inspection do not also give evidence of COPLs of a different kind. In contrast to Maltese, the inflectional paradigm of nouns in Estonian and Georgian additionally involves case distinctions which render the picture even more colourful.

3.1 Estonian

The Uralic language Estonian allows for the competition of several strategies of pluralization in the domain of nouns. The so-called inner and outer local cases for instance are involved in a pattern of variation which embraces three options. The plural can be expressed either by the suffix *-de/-te* or by the suffix *-i*. The choice of the plural marker has repercussions on the segmental chain of the stem. This can be shown for the noun *hammas* 'tooth' (Viitso 2003: 39). The stem is *hammas* with *-te* but *hamba* with *-i*. A dedicated plural marker is absent from the third option because this time the case suffixes are directly attached to a special plural stem, viz. *hambu-*. Table 2 illustrates this situation for the local cases.

Table 2: COPLs in Estonian (local cases).

category	*de*-plural			*i*-plural			stem plural	
	stem	**plural**	**case**	**stem**	**plural**	**case**	**stem**	**case**
illative₂			-sse			-sse		-sse
inessive			-s			-s		-s
elative	hammas	-te	-st	hamba	-i	-st	hambu	-st
allative			-le			-le		-le
adessive			-l			-l		-l
ablative			-lt			-lt		-lt

Blevins (2008: 258–261) discusses different prosodically oriented approaches to allomorphy in Estonian declension only to rebut them altogether. The plethora of options is there for purely morphological reasons (Blevins 2008: 264). Meaning: a valid functional explanation in terms of a division of labour – be it pragmatically or semantically – has not been put forward yet. The ternary set of COPLs *hammas-te-* ~ *hamba-i-* ~ *hambu-* thus counts as "meaningless" in the sense that speakers of Estonian do not opt for a given plural with a view of conveying a message that could not also be transferred by the other pluralization strategies. Note that the three options differ widely in terms of their token frequency. According to the results of a search on Estonian Web 2019 (etTenTen 19) (visited 13 September, 2021), the special plural stem *hambu-* is unattested in combination with the above six local cases. There are altogether 864 tokens of pluralized local cases of *hammas* 'tooth' 854 (~ 99%) of which involve the *te*-plural whereas only 10 matches (~ 1%) could be found for the *i*-plural of the same lexeme. Checking informally the frequencies of selected word-forms of a single lexeme is certainly not enough to settle the issue but the data are nevertheless suggestive of unbalanced overabundance in the sense that there must be conditions which determine the choice of the minority options.

3.2 Georgian

In the Kartvelian language Georgian, nouns are generally pluralized by way of adding the suffix *-eb* to the stem as e.g. NOM.SG *monadire* 'hunter' → NOM.PL *monadire-eb-i*. Beside the *eb*-plural, there is, however also the *n*-plural. In principle, each noun can be pluralized in both ways, i.e., NOM.PL *monadire-n-i* is possible, too. Yet, the two options differ from each other as to several parameters as results from the comparison of the two plural paradigms of the noun *glexi* 'farmer' in Table 3 (Fähnrich 1986: 47 and 51).

Table 3: *eb*-plural vs *n*-plural in modern Georgian.

category	eb-plural			n-plural		
	stem	plural	case	stem	plural	case
nominative			-i		-n	-i
ergative			-ma			-t(a)
genitive			-is(a)	glex		-t(a)
dative	glex	-eb	-s(a)			-t(a)
instrumental			-it(a)			
adverbial			-ad(a)			
vocative			-o	glex	-n	-o

In contrast to the *eb*-plural, the *n*-plural is defective in the sense that there are no forms for the instrumental and the adverbial. Moreover, there is also syncretism since genitive, dative, and ergative are expressed identically. Furthermore, the syncretic marker *-t(a)* is also described as a portmanteau morph which encodes both number and case whereas there are no comparable portmanteaus in the *eb*-plural. Tschenkéli (1958: 60–61) describes the *n*-plural as residual category ("alter Plural" = 'old plural') which is attested mostly in poetry (but hardly ever in prose). In addition, the genitive of the *n*-plural is relatively frequent with attributive nouns in administrative, technical, and political texts. Therefore, the distinction of *eb*-plural and *n*-plural in Georgian is not as "meaningless" as that of the three pluralization strategies in Estonian since there is a relatively strong association of the *n*-plural with certain styles, registers, and genres (Aronson 1990: 118 and 147). The *eb-n* opposition does not help to distinguish meanings but it is sensitive to sociolinguistic parameters. Thus, the *n*-plural survives in contemporary Georgian in stylistic niches.

According to Fähnrich (1994: 55), the *n*-plural was still common in the Old Georgian period when it already had to compete with the *eb*-plural which later on ousted its contestant almost completely in the spoken language. On early stages of Georgian, the case paradigm of the *n*-plural also featured (syncretic) forms of the instrumental and adverbial which were expressed by the above portmanteau morph *-t(a)*. It is assumed that the *n*-plural arose from a former dual in the period before the earliest written records in Georgian.

The coexistence of the two plurals is an old phenomenon in Georgian. We are witnessing the slow withdrawal of the *n*-plural from the grammatical system. Presently, it is still available as a stylistically marked alternative to the dominant *eb*-plural. The two cellmates do not enjoy equal rights. It is wrong, however, to declare the *n*-plural obsolete because it still has a function of its own albeit not a properly grammatical one since it has become a stylistic device. Mutatis mutandis this situation is reminiscent of the co-existence of *vo* and *vado* 'I go' (= 1SG of *andare* 'go') in Italian as described in Thornton (2013).

3.3 Nahuatl

Launey (2011: 19–21) summarizes the basics of plural marking in Colonial Nahuatl (Uto-Aztecan) of the 16th-19th century as follows:

> It is not possible to predict the plural of a noun from its singular form. This constitutes the only real difficulty in the morphology of Nahuatl grammar. [] The plural in Nahuatl was never very fixed, and there are many doublets. [] In doubtful cases, adding *-mê* after vowels and *-tin* after consonants is an acceptable strategy. [] *Only the names of animate beings [] can be put in the plural* [original italics]. [] A very small number of nouns for inanimate 'things' have a plural form, probably as a result of mythological personification.

The animacy-based system of Colonial Nahuatl excludes the vast majority of inanimate nouns from pluralization. The regular patterns for pluralizing animate nouns are given in Table 4 according to Launey (2011: 19–21).

Table 4: Patterns of pluralization in Colonial Nahuatl.

pattern	singular	plural	meaning
V]-*tl* > -*ʔ*	*cihuā-tl*	*cihua-ʔ*	'woman'
reduplication + -*tl* > -*ʔ*	*teō-tl*	*tē~teo-ʔ*	'god'
C]-*tli* > -*tin*	*oquich-tli*	*oquich-tin*	'man'
reduplication + C]-*tli* > -*tin*	*tōch-tli*	*tō~tōch-tin*	'rabbit'
-*Ø* ~ -*in*] > -*meʔ*	*mich-in*	*mich-meʔ*	'fish'
	chichi	*chichi-meʔ*	'dog'

Deviations from these patterns are relatively common. Launey (2011) mentions pairs of COPLs like *cihua-ʔ* ~ *cihuā-meʔ* 'women', *cō~cōhua-ʔ* ~ *cōhuā-meʔ* 'snakes', *oquich-tin* ~ *oquich-meʔ* 'men', *mich-tin* ~ *mī~mich-tin* 'fish(es)', etc. which do not seem to reflect any meaning differences between the members of a given COPL-pair. In the beginning of the written documentation of Colonial Nahuatl in the second half of the 16th century, the competition between the alternative plurals was not particularly strong. On the basis of Eisinger's (1998) computerized index of the monumental Codex Florentinus, I have found evidence of COPLs only for *mich-tin* 'fish' (one token) vs *mi~mich-tin* 'fish' (20 tokens) and *cō~cōhua-ʔ* 'snakes' (27 tokens) vs *cōhuā-mê* 'snakes' (two tokens).

Massive borrowing from Spanish has contributed to the reshaping of the original system. The erstwhile exclusion of inanimate nouns from pluralization was

given up while the Spanish plural marker -s became obligatory with Spanish loanwords (Olko, Borges and Sullivan 2018: 478–482). In the extended transition period,
loanwords could be pluralized in three ways, namely by retaining the Spanish pluralizer -s or by replacing it with the indigenous pluralizer -me? or by combining
both markers so that triplets like SG *cavallo* 'horse' → PL *cavallo-s* ~ *cavallo-me?* ~
cavallo-s-me? (Eisinger 1998: II, 98) arose temporarily. Meaning differences are not
reported in connection to the members of this set of (ephemeral) COPLS.

Apart from the relaxation or complete abolition of the erstwhile animacy-
based restrictions of pluralization, the modern varieties of Nahuatl do not behave
homogeneously as to plural marking. For the Nahuatl of Mecayapan (Veracruz),
Wolgemuth (1981: 45–46) postulates the existence of five patterns of pluralization
which are featured in Table 5. Grey shading highlights those cells which host the
inherited plural suffix -*mej* (= Colonial Nahuatl -*me?*).

Table 5: Patterns of pluralization in Nahuatl (Mecayapan).

pattern	singular	plural		meaning
-mej	cajli	cajli-mej		'house'
∈]-mej	ama'	ama-mej		'paper'
-quej ~ -mej	huehuej	huehuet-quej	huehuet-mej	'elder'
-quej	sihua'	sihuat-quej		'woman'
reduplication	taga'	taj~taga'		'man'

Three of the five patterns cover sizable groups of nouns. COPLS, however, occur only
with the noun *huehuej* 'elder', namely *huehuet-quej* and *huehuet-mej*. The suffix
-*mej* is said to have the widest distribution over nouns in this variety of Nahuatl.
In contrast, -*quej* is attested only with three animate nouns. What we witness is
probably the pre-final stage of the generalization of the pluralizer -*mej* which is
making inroads into the erstwhile presumably much larger domain of -*quej*. The
COPLS *huehuet-quej* ~ *huehuet-mej* reflect the spread of -*mej* beyond the boundaries
of its former domain. Since Wolgemuth (1981: 45) does not mention any meaning
differences between the two plurals, I assume that they are fully synonymous.

3.4 Breton

Since several contributions to this edited volume address number marking in
Welsh whose diachrony and typology is expertly described in Nurmio (2019), I
discuss only data from a close relative of Welsh, namely Breton (cf. also Acquaviva

2008: 234–265).[15] Trepos (1982) has dedicated a monograph to the plural in Breton. This number category is depicted as a multifaceted phenomenon with an enormous range of variation. On account of Trepos's dialectological methodology, the results of his study must be carefully examined because many of the supposed COPLs turn out to be combinations of the choices of different regional varieties of Breton. In the standard, we find SG *bugel* 'child' → PL *bugale* ~ *bugalez* whereas in spoken (regional) Breton PL *bugaled* and PL *begali* are attested, too (Trepos 1982: 58–59). It remains unclear from Trepos's description whether all four plurals cooccur in one and the same variety.

However, even though the extent of dialectal differences is often hard to pin down on the basis of the extant literature, there remain sufficient cases of uncontroversial COPLs to be taken note of. Trepos (1982: 77–84) reviews a selection of COPL-pairs to determine the conditions which regulate the choice of plural. In most of the cases, the different plurals reflect meaning differences such as SG *gwaz* 'husband; client' → PL *gwazed* ~ *gwersed* 'husbands' ≠ *gwizien* 'clients.'[16] In addition to these semantically loaded cases, there are also those for which neither dialectal differences or meaning distinctions can be invoked. Favereau (1997: 33) mentions SG *prenest(r)* 'window' → PL *prenester* ~ *prenistri* ~ *prenest(r)où* which can be used interchangeably. Half a dozen plurals are given by Helias (1986: 217) for SG *marc'h* 'horse' → PL *kezeg* ~ *mirc'hed* ~ *mirc'hien* ~ *mirc'hier* ~ *merc'h* ~ *mirc'hi*[17] to which Favereau (1997: 34) adds PL *marc'hed* ~ *marc'hoù* with the latter being used predominantly for sawhorses. Note the suppletive *kezeg* 'horses' which is probably the odd one out in this long list of plurals as it seems to invite an interpretation as collective whose preferred singulatives are *loen-kezeg* ~ *penn-kezeg* ~ *jao* ~ *roñse* rather than *marc'h* (Favereau 1997: 34).

In Table 6, I present a small selection of COPLs mentioned by Favereau (1997: 33–44). These cases have no specific dialectal background. Plural marking via internal vowel change alone is considered an archaism whereas pluralization via

---

15 COPLs seem to be a characteristic trait of the Brythonic branch of the Celtic languages. In Goidelic, evidence of COPLs is not unknown either (Acquaviva 2008: 162-194). Alternative forms occur relatively frequently in the genitive plural, but I register these alternations as case allomorphy with no connection to number marking. In the case of Scots-Gaelic SG *deoch* 'drink' → PL *deochan* ~ *deochannan* (Mark 2004: 646) the alternative forms count for the entire paradigm of the plural and may thus be accepted as COPLs. In Calder's (1980) historicizing grammar of Scots-Gaelic – originally published in 1923 – COPLs are registered in sizable numbers such as SG *biadh* 'food' → PL *bìdh* ~ *bidhe*, SG *spong* 'sponge' → PL *spuing* ~ *spogan*, SG *tom* 'round hillock' → PL *tuim* ~ *toman* (Calder 1980: 83).

16 I have modified Trepos's examples orthographically according to the regulations of the modern standard.

17 I have modified Helias's examples orthographically according to the regulations of the modern standard.

suffixation of -*où* represents the most productive mechanism of plural marking in modern Breton.

Table 6: Selected Breton COPLs without meaning differences.

singular	meaning	plural 1	plural 2	plural 3
bro	'country'	*broioù*	*broezioù*	
buoc'h	'cow'	*bioù*	*buoc'hed*	
gavr	'goat'	*gevr*	*garvred*	*givri*
godell	'pocket'	*godilli*	*godelloù*	
kastell	'castle'	*kestell*	*kastelloù*	*kistilli*
kog	'cock'	*kèjer*	*kigi*	*kogoù*
mantell	'overcoat'	*mantilli ~ mintilli*	*mantelloù*	*mentell*
maout	'ram'	*meot*	*maoutoù*	*maouted*
moteur	'motor'	*moteurien*	*motorioù*	
sparfell	'sparrowhawk'	*sparfelled*	*sparfilli*	

Wherever one of the alternatives is a plural form ending in -*où* we are probably facing an innovation, i.e. the *où*-plural is gaining ground on the other strategies whose domain seems to be shrinking. The dynamics of the *où*-plural notwithstanding, its competitors have not yet been ousted completely so that many of the COPL-sets remain stable at least for the time being.

3.5 Dutch (and Afrikaans)

Van den Toorn (1977: 156–157) characterizes the suffixes -*en* and -*s* as the most frequent plural markers of Dutch nouns alongside the unproductive -*eren* and the Latin suffixes -*i*, -*a*, and -*es* on loanwords. According to the author, the *en*-plural is considered the unmarked case of pluralization by native speakers of Dutch. The domains of -*en* and -*s* are shaped by a network of factors among which we find phonological (including prosodic), morphological, semantic, and etymological criteria. What is important for the topic of this paper is the fact that "[b]ij enige woorden kan zowel -*en* als -*s* gebruikt worden zonder dat er verschil in betekenis optreedt" ['with some words both -*en* and -*s* can be used without triggering meaning differences'] (Van den Toorn 1977: 156). In other cases, the *en*-plural and *s*-plural of a noun correlate with different meanings such as SG *patroon* 'patron; pattern' → PL *patronen* 'patterns' ≠ PL *patroons* 'patrons'. Acquaviva (2008: 36) addresses Dutch COPLs too and concludes that (a) "there is no one-to-one relation between irregular form and irregular meaning" and (b) "the semantic distance between the two plural forms varies considerably".

Geerts et al. (1984: 66–68) devote a chapter of their normative grammar of Dutch to the problem of "meaningless" COPLs. To their mind, the competition of *en*-plural and *s*-plural is largely a matter of style. The plural in -*en* is often (but not always) considered to represent the more refined stylistic level. The preference for -*en* in the context of high style is, however, not shared by all speakers throughout the Dutch speech territory. Moreover, some *en*-plurals are classified as archaic or typical of the written register. The authors provide a sizable list of COPLs from which I pick ten pairs to compare the token frequencies of their members which are disclosed in Table 7. The frequencies result from a search on Dutch Web 2014 (nlTenTen 14) carried out on 17 September, 2021. The nouns are ordered top-down according to the decreasing token frequency of the *en*-plural. Grey shading marks out those cases which give evidence of a majority of *s*-plurals.

Table 7: Dutch COPLs and token frequencies.

singular	meaning	*en*-plural	tokens	*s*-plural	tokens
gemeente	'congregation'	gemeenten	230,401	gemeentes	16,580
ziekte	'illness'	ziekten	41,768	ziektes	25,182
olie	'oil'	oliën	41,269	olies	234
vitamine	'vitamin'	vitaminen	22,892	vitamines	18,404
ventilator	'ventilator'	ventilatoren	5,187	ventilators	312
dienaar	'servant'	dienaren	4,640	dienaars	756
kolonie	'colony'	koloniën	2,923	kolonies	3,177
appel	'apple'	appelen	2,618	appels	23,157
lector	'lecturer'	lectoren	1,871	lectors	4
harmonie	'harmony'	harmoniën	64	harmonies	106

Seven out of ten COPL-pairs prefer the *en*-plural over the *s*-plural. The range of the token frequencies is wide. Figure 2 reveals that the ratios between the two options are relatively close for eight of the COPL-pairs.

For the COPLs *oliën-olies* the ratio is 176:1 whereas that of *lectoren-lectors* is 468:1. The quantitative discrepancy is such that these cases can be considered to represent unbalanced overabundance. Geerts et al. (1984: 67) argue that

[v]ooral bij benamingen van hoge maatschappelijke functies als *professor, lector, rector* wordt het meervoud op -*s* als stilistisch niet adequaat beschouwd.[18]

---

18 My translation: "especially with denominations for higher societal functions such as *professor, lector, rector* the plural in -*s* is considered to be stylistically inadequate."

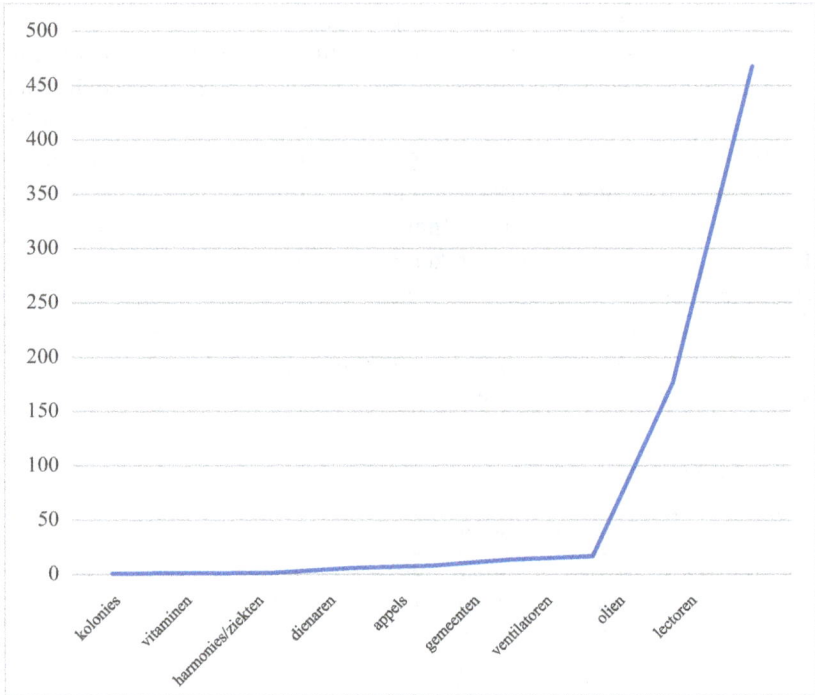

Figure 2: Ratios of *en*-plurals and *s*-plurals in Dutch.

The overwhelming preference for *lectoren* as opposed to *lectors* is thus motivated by considerations of style. This explanation cannot, however, be applied to *oliën-olies*. The closest one can get to balanced overabundance is the pair *kolonies-koloniën* with the ratio 1.08:1 followed by *vitaminen-vitamines* (ratio: 1.24:1), *harmonies-harmoniën* and *ziekten-ziektes* – both with the ratio 1.65:1. To complement this impressionistic review of the frequencies of competing plurals, I add two instances from learned vocabulary, namely SG *museums* 'museum' → PL *musea* (49,638 hits) ~ *museums* (1,152 hits) with the ratio 43:1 and SG *crisis* 'crisis' → PL *crises* (4,989 hits) ~ *crisen* (6 hits) with the ratio 831,5:1. In these cases, the preference for the "foreign" pluralizers -*a* and -*es* can probably be explained again with reference to stylistic aptness.

It is worth noting that Afrikaans, the African offspring of Dutch, also gives ample evidence of COPLs which involve the pluralizers -*s* and -*e* (< -*en*) as well as the Latinate plurals (Donaldson 1993: 80–84). There are cases like SG *plafon* 'ceiling' → PL *plafonne* ~ *plafons*, SG *tekening* 'drawing' → PL *tekenninge* ~ *tekennigs*, SG *hotel* 'hotel' → PL *hotelle* ~ *hotels*, SG *sentrum* 'centre' → PL *sentra* ~ *sentrums*, SG *akademikus* 'academic' → PL *academici* ~ *akademikusse*, etc. In their vast majority, Afrikaans

COPLs do not reflect meaning differences. With reference to nouns ending in *-ing*, Donaldson (1993: 80) argues that

> [t]heoretically all nouns [of this type] can take either ending, [*-e*] being more frequent in formal spoken or written style. In practice, however, several of these nouns are never found with an *-s* plural, because they belong by definition to higher style.

Some archaic *e*-plurals are typical of formal religious contexts. The plural *kerne* 'nuclei' (← SG *kern* 'nucleus; kernel') is considered "more formal and scientific than *kerns*" (Donaldson 1993: 83). The learned *i*-plural of nouns (misleadingly labelled *ci*-plural) ending in *-us* is presented as more common than the *e*-plural (Donaldson 1993: 84). Thus, in Afrikaans, the choice of plural from a COPL-pair is overwhelmingly a matter of style.

3.6 Intermediate summary

The five previous case-studies and the additional data discussed at the beginning of Section 3 prove first and foremost that cross-linguistically COPLs are not as marginal as one might think. The phenomenon is attested in many languages which belong to different language families, are spoken in different parts of the world, and display significant typological differences among each other. Therefore, finding evidence of COPLs in a given languages should no longer surprise us. Furthermore, it has come to the fore that languages tolerate COPLs although the competing plurals do not distinguish lexical or grammatical meanings. What frequently seems to justify the co-existence of several plurals of the same noun is style. The different options are made use of according to the requirements of a communication setting – the distinctions of formal vs colloquial and written vs spoken are often invoked to demarcate the domains of the competing wordforms. The plurals which meet the criteria imposed by formal style and written register are frequently characterized as archaic. Innovative plurals are sometimes borrowed from another language. Accordingly, at least some pairs of COPLs can be understood as a momentary phase in an ongoing process of language change which might result in the ousting of one of the competitors. The token frequencies of the COPLs are by far too diverse to allow for sweeping generalizations. Since the available sources are not always explicit as to the distribution of alternative plurals, I do not want to jump to conclusions not the least because the inventory of the nouns which take COPLs has not been determined yet for most of the languages discussed in this paper. Only if and when we know how many and which nouns are affected by the phenomenon under scrutiny will we be in a position to evaluate COPLs thoroughly and to the benefit of our discipline. The idea that the empirical part of the project urgently calls for being given more attention receives support from the data discussed in the subsequent Section 4.

4 "Meaningful" COPLS

As with "meaningless" COPLS, the interpretation of the data is at times demanding. I highlight some of the usual problems first in relation to grammatical issues (Sections 4.1–4.2); lexical matters are discussed further below in Sections 4.3–4.5.

4.1 Animacy and sundry categories

Since the category that is at issue in this study is number, one expects that COPLS which are grammatically meaningful give rise to an additional value that belongs to the same category. However, cases like Cambodian SG *kamoh* 'mistake' → (general) PL *kamohniania* 'mistakes' ≠ (distributive) PL *kamohpsee:ngpsee:ng* 'various mistakes' (Haiman 2011: 50) do not meet the necessary criteria for being discussed in the wider context of overabundance and related phenomena because both the general and the distributive plural are in principle legitimate on each pluralizable noun. Language-internally their presence does not constitute a morphological mismatch. Overabundance as an analytic concept makes sense only if there is variation in a given language in the sense that only certain nouns or cells of a paradigm allow for more distinctions than other nouns or cells of the same paradigm. Moreover, Corbett (2000: 117) argues that "though there are similarities between distributives and number values, distributives are not a part of the number system." I come back to this issue in connection with the discussion of collectives in Section 4.2.

Paradigms can be multidimensional in the sense that in addition to the different values of the category of number which have to be distinguished formally there are also those of other categories such as case, definiteness, gender, and/or animacy. To save space, I focus only on the latter. In many languages worldwide, animacy (Corbett 2000: 55–56) plays a role for the application of pluralization strategies.[19] This holds for instance, for Tajiki (Indo-Iranian). Similar to the Persian case

---

19 Nurmio (2019: 40-45 and 235-239) shows that animacy is also relevant for other number categories such as the dual and in the domain of numeratives.

in Section 3, Tajiki gives evidence of borrowed Arabic plurals (including broken plurals). In the inherited part of the lexicon, a distinction is made between two kinds of pluralization, namely (a) via the suffix *-ho* which is licit with all kinds of stems and (b) the suffix *-on* which is restricted to animate nouns and a closed set of other nouns. Since the use of *-ho* is unrestricted we find situations like SG *mard* 'man' → PL *mardon* ~ *mardho* (Khojayori and Thompson 2009: 19). Rzehak (2019: 31 footnote 38) elaborates on the COPL-patterns as follows. The choice of pluralizer is stylistically motivated. In high style, *-on* is preferred if reference is made to highly esteemed persons. This means that the use of *-on* is also an indirect means of expressing politeness. Politeness belongs to pragmatics. The distinction animate vs inanimate is of grammatical relevance. However, the coexistence of the COPLs *mardon* and *mardho* does not result in the creation of an additional number value. On the other hand, in Tajiki, COPLs can occur only in the domain of animate nouns (and those nouns which associate with this category without being properly animate).

In the Austronesian language Chamorro, there are three classes of nouns (Chung 2020: 109–112):
(a) (certain) human nouns which boast special plural forms (obligatory), e.g. SG *låhi* 'man' → PL *la~låhi*,
(b) (mostly) human nouns which optionally take the plural prefix *man-*, e.g. SG *ma'estra* 'female teacher' → PL *man-ma'estra*, and
(c) non-human nouns which are optionally pluralized by the post-nominal pronoun *siha* 'they' (preferably if the NP is definite), e.g. SG *palåbra* 'word' → PL *(i_DEF) palåbra siha*.

In (3), I present two sentences from the same story told by the same native speaker in which the animate noun *ma'estra* 'female teacher' (< Spanish *maestra*) is pluralized in two different ways.

(3) Chamorro
(a) prefix (Taimanglo, Yamashita and Rivera 1999: 54)
Annai måtto si Juan gi sigiente diha, mandañña' i
when arrive DEF Juan in following day PL:unite DEF
***man-ma'estra** gi me'nan un tenda*
PL-TEACHER:F in in_front_of INDEF tent
'When Juan arrived the next day, the **female teachers** had gathered in front of a tent.'

(b) postnominal marker (Taimanglo, Yamashita and Rivera 1999: 55)

Gi	sanhalom	i	didide'	na	tiempo,	ha		kombida	si
in	DIR:inside	DEF	a_little	LK	time		3SG.ERG	invite	DEF

Juan	i	**ma'estra**	**siha**	para	u	ma	li'e'
Juan	DEF	**teacher:F**	**PL**	for	IRR	3PL.ERG	see

i	adilånto.
DEF	improvement

'After a short time, Juan invited the **female teachers** to see the improvement.'

The competing pluralization strategies for animate nouns of the type *ma'estra* are so different formally that one hesitates to speak of cellmates unless we accept periphrasis as a legitimate strategy within an otherwise inflectional paradigm. This difficulty notwithstanding, we are facing competition in connection with the expression of the grammatical category of number. Exactly what determines the choice of construction remains an open question.[20] The different syntactic relations of subject in (3a) and object in (3b) are irrelevant for the distribution of the two options of pluralization. What can be said nevertheless is that similar to the previous case from Tajiki, COPLs occur only in the class of human nouns, i.e. with those nouns which occupy the top rank on the animacy hierarchy. As in Tajiki, no additional number value emerges in the context of these COPLs.

According to the line of argumentation adopted by Thornton (2010: 21–25), the failure of the COPLS to yield an additional number value is tantamount to the rebuttal of overdifferentiation. In Tajiki and Chamorro, the COPLs are meaningful on the dimension of animacy but not for the category of number. In Section 4.2, a different scenario is scrutinized which involves a category that is related to the category of number but not part of it.[21]

4.2 Lower Sorbian

In the West Slavic minority language Lower Sorbian, nouns morphologically distinguish the numbers singular, dual, and plural. Nouns also inflect for case. There are several inflection classes and a ternary set of genders (masculine, fem-

---

20 Isabel Compes has suggested that pragmatic focus might be a factor. Whether this is indeed the case cannot be determined on the basis of my corpus which contains only written texts.
21 Another case of meaningful exploitation of number marking was mentioned to me by Paul Widmer and Jürg Fleischer. In Yiddish, it is possible to use the Hebrew dual suffix to form a plural which has evaluative connotations (Jacobs 2005: 163).

inine, and neuter). In the nominative plural of certain human nouns, two kinds of plurals are attested, namely the expected "regular" plurals in *-y/-e* and plurals marked by *-i/-(j)e/-(a)*. Traditionally, the regular plurals refer to multitudes of individuals whereas the additional plurals receive an interpretation as collective (Janaš 1976: 112–113). The distinction between the COPLs is restricted to the nominative. Animacy is crucial since only human nouns are subject to this differentiation. All nouns displaying COPLs are masculine. The class is closed and counts fourteen members which are surveyed in Table 8. The grey shaded columns contain the regular plural forms.

Table 8: Masculine COPLs in Lower Sorbian.

singular	meaning	plural 1	plural 2	plural 3	plural 4
bratš	'brother'	*bratšy*	*bratśi*		
bur	'farmer'	*bury*	*burje*	*burja*	
cart	'devil'	*carty*	*carti*		
Cygan	'gypsy'	*Cygany*	*Cygani*	*Cyganje*	
forman	'coachman'	*formany*	*formani*		
kmotš	'godfather'	*kmotšy*	*kmotśi*	*kmotśe*	*kmotśa*
knecht	'farmhand'	*knechty*	*knechśi*		
kněz	'master'			*kněze*	*kněža*
kśesćijan	'Christian'	*kśesćijany*	*kśesćijani*	*kśesćijanje*	
nan	'father'	*nany*	*nani*		
sused	'neighbour'	*susedy*	*suseźi*		
šołta	'mayor'	*šołty*	*šołtśi*		
tatań	'heathen'		*tatani*	*tatanje*	
Žyd	'Jew'	*Žydy*	*Žyźi*		

Ten nouns display two COPLs, three nouns are invested with three COPLs each, whereas *kmotš* 'godfather' allows for four COPLs. Table 8 hosts three borrowings from (Low) German, namely *bur* 'farmer' (< Low German *Buer* 'farmer'), *forman* 'coachman' (< Low German *Fohrmann* 'coachman'), and *knecht* 'farmhand' (< (Low) German *Knecht* 'farmhand'). Historically these borrowings have to be considered innovations. It is interesting that these latecomers to the Lower Sorbian lexicon are subject to the distinction of two plurals, meaning: at the time of their integration into the replica language's system the distinction must have been productive. Impressionistically the semantics of the nouns in Table 8 invoke the picture of a pre-industrial society. Twelve of the fourteen nouns attest to the coexistence of *y*-plural and *i*-plural. If a noun takes the *y*-plural it always also has an *i*-plural. Only *kněz* 'master' lacks the *i*-plural.

Without exception, the different COPLs trigger normal plural agreement on targets as in the NPs *wšak-e knĕz-e* 'some masters' and *naš-e nan-i* 'our fathers' (Janaš 1976: 113) where the quantifier and the possessive pronoun reflect the *e*-plural independent of the choice of plural of the head noun. The two examples in (4) illustrate the occurrence of the COPLs in different sentential contexts. Note that in (4b) the nominative plural is used in the function of the vocative.

(4) Lower Sorbian (Janaš 1976: 113)
    (a) Plural 1
        *Moje   tśo   starše   **bratš-y**   su   mĕ*
        POSS.1SG:PL three old:CMPR:PL **brother-PL** AUX.3PL 1SG.DAT
        kuždy nĕco darili.
        each something give:PART:PL
        'My three elder **brothers** have each given me something.'
    (b) Plural 2
        *Cesćone **bratś-i** a sotše!*
        respect **brother-PL** and sister:PL
        'Show respect, **brothers** and sisters!'

According to the usual analysis of the Lower Sorbian examples, *bratšy* 'brothers' in sentence (4a) refers to three individuals whereas *bratśi* 'brothers' in sentence (4b) refers to brothers collectively. This supposed collective forms the first conjunct of a coordinative construction whose second conjunct is a regular (feminine) plural, i.e. collective and plural are semantically compatible with each other.

Like the distributive mentioned in Section 4.1, the collective is not accepted as a value of the number category by Corbett (2000: 117–118) whose lead Thornton (2010: 26–28) follows. Since the collective is counted out as a number value, the Lower Sorbian case and those cases which resemble it do not constitute instances of overdifferentiation. The dialectics of the argumentation against treating the collective as number value are intricate especially because the conceptual affinity to plurality is not denied by the above authors. Granted that overdifferentiation does not apply, Lower Sorbian gives evidence of COPLs which in turn instantiate semantically conditioned overabundance. For speakers of Lower Sorbian, the choice of the plural form is meaningful. The picture is rendered complicated by the presence of more than two options which is the case for four of the nouns in Table 8. We are perhaps facing two interacting kinds of overabundance, namely the semantically conditioned overabundance (plural vs collective) in combination with unconditioned overabundance if two regular plurals and/or two collectives are available.

Lower Sorbian offers further data which guide us to the topic of Section 4.3. Two neuter nouns are equipped with COPLs, namely SG *woko* 'eye' → PL *wocy* 'eyes' ≠

PL *woka* 'eye-shaped things (e.g. ear of a needle, sling)' and SG *wucho* 'ear; handle' →
PL *wušy* 'ears' ≠ PL *wucha* 'handles' (Janaš 1976: 113). The meanings of nouns which
are polysemous or homonymous in the singular are differentiated formally in the
plural. This strategy is applied frequently by numerous languages. Some of the analytical problems which are connected to this strategy are discussed in Section 4.3. In
Section 4.4–4.5, data from Romanian and Albanian are presented.

4.3 Lexical meanings

Several of the previously discussed languages give evidence of "meaningless" and
"meaningful" COPLS. Grammatically "meaningful" COPLS which fulfil the criteria of
overabundance are hard to come by. The picture is different when we consider the
differentiation of lexical meanings by way of pluralizing a given noun in different
ways. At this point we enter the territory of Acquaviva's (2008) lexical plurals. Apart
from some heterographic plurals, French uses this strategy in only three cases, namely
SG *ciel* 'sky, heaven' → PL *cieux* 'skies, heavens' ≠ PL *ciels* 'bed canopies', SG *œil* 'eye' →
PL *yeux* 'eyes' ≠ PL *œils* 'hole (etc.)', and SG *travail* 'work' → PL *travaux* 'works, efforts'
≠ PL *travails* 'apparatus employed for veterinarian operations on quadrupeds' (Grevisse 1964: 220–223). In the singular, the different meanings associated with a given
noun can be distinguished only on the basis of contextual information. In the plural,
however, the different meanings are represented by formally distinct wordforms. One
of the COPLS is typical of a language for special purposes. This also holds for SG *ail*
'garlic' → PL *aulx* ~ *ails* where meaning distinctions are not at issue. Grevisse (1964: 221)
assumes that *ails* belongs to the special language of botany whereas *aulx* is in general
use. Moreover, the COPLS *travaux* and *travails* result from the merger of two lexemes
(Latin *trepalium* 'torturing device' and the Old French verb *travailler* '(to) work') in the
singular whose distinct plurals survived into the present (Grevisse 1964: 223).

This takes us to the vexed question of how many nouns there are. If we assume
polysemy for the singular, there is one noun with several plurals. In contrast, if
homonymy is assumed, there are no COPLS at all since there are several incidentally
homophonous singulars, i.e. distinct nouns. Thornton (2010: 37–39) discusses this
serious problem at length to show that it is almost impossible to decide satisfacto-
rily whether polysemy or homonymy applies. The French case *travaux* vs *travails*
can be traced back etymologically to two different sources so that an interpretation
as homonymy seems to be plausible. As to the COPLS *cieux-ciels* and *yeux-œils*, polysemy offers the better solution since the meanings of the different plurals can be
connected to each other via metonymy and/or metaphor. Since etymological information is available only for historically well-documented languages, the diachronic
development of COPLS of many languages remains inaccessible to us. The descrip-

tive grammars of the languages reviewed in the two subsequent sections clearly distinguish polysemous from homophonous cases. To avoid comparing like with unlike, I only take those nouns into consideration which are explicitly registered under the rubric of polysemy. Moreover, I adopt Thornton's (2010: 38–39) Solomonic decision not to take a definitive stance in cases of doubt.

Sections 4.4–4.5 feature two languages which give ample evidence of COPLs. In both cases, there is a plethora of nouns which attest to semantically unconditioned overabundance alongside numerous instances of semantically conditioned overabundance. Both types of overabundance are sketched in the subsequent sections to provide the reader with an at least superficial impression of the prominent position COPLs have in the systems under review. The two language – Romanian and Albanian – belong to different branches of the Indo-European language family, viz. Romance and (the internal isolate) Albanian. The languages are spoken in each other's vicinity in the Balkans where they are members of the Balkan *Sprachbund*. Romanian and Albanian nouns inflect for case. It is therefore hardly surprising that these object languages also share properties in the domain of COPLs.

4.4 Romanian

Beyrer, Bochmann and Bronsert (1987: 82) characterize the phenomenon of interest to us as follows:

> Die im Rum[änischen] ziemlich häufig auftretenden Mehrfachformen im Plural beruhen auf phonetischen, morphologischen und semantischen Ursachen. Es handelt sich im wesentlichen um soziolinguistisch differenzierte Prozesse[22]

The authors distinguish between formal and colloquial plurals. In Table 9, ten nouns illustrate this kind of COPLs. The nouns are ordered top-down according to the decreasing token frequency of plural 1. The frequencies have been determined by way of searching the Romanian Web 2016 (roTenTen16) (last visited on 13 December, 2021). The grey shaded cells host options which the reference grammar qualifies as infrequent or archaic.

All the nouns featured in Table 9 are attested in the corpus so that it is possible to compare the token frequencies for the pairs of COPLs. Figure 3 reveals that the ratios between the two options differ widely across the nouns.

---

22 My translation: "The frequently attested multiple forms of the plural in Romanian are based on phonetic, morphological, and semantic factors. In principle, these are socio-linguistically differentiated processes."

Table 9: Register-dependent COPLs in Romanian.

singular	meaning	plural 1 (formal)	tokens	plural 2 (colloquial)	tokens
hotel	'hotel'	hoteluri	57,579	hotele	824
francez	'Frenchman'	francezi	34,522	franceji	66
simbol	'symbol'	simboluri	24,478	simboale	50
proverb	'proverb'	proverbe	3,704	proverburi	28
copertă	'envelope'	coperte	1,705	coperți	1,479
otravă	'poison'	otrăvuri	889	otrăvi	3,052
burghez	'bourgeois'	burghezi	795	burgheji	110
cratiță	'saucepan'	cratițe	270	crătiți	55
zeamă	'broth, juice'	zemuri	160	zemi	18
jalbă	'complaint'	jalbe	144	jălbi	29

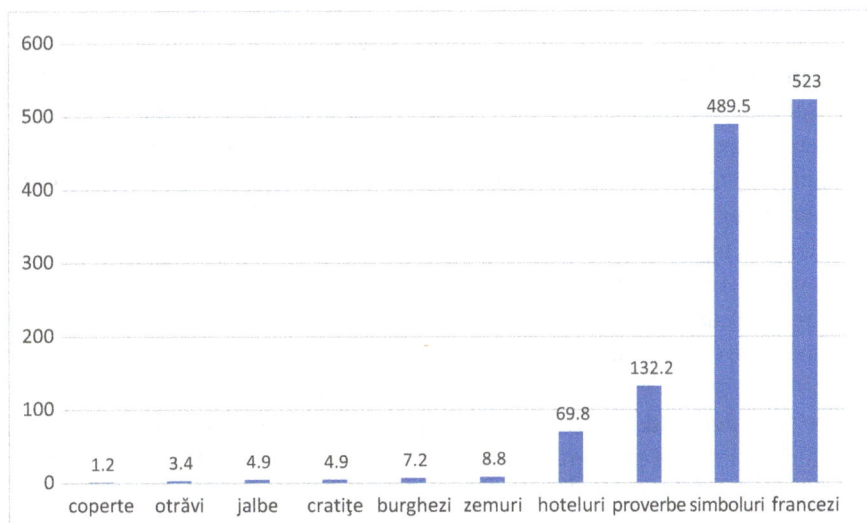

Figure 3: Ratios of formal and colloquial plurals in Romanian.

For the nouns *copertă* 'envelope', *otravă* 'poison', *jalbă* 'complaint', *cratiță* 'saucepan', *burghez* 'bourgeois', and *zeamă* 'broth, juice' the ratios range from 1.2:1 to 8.8:1 whereas the remaining four nouns in Figure 3 cover ratios from 69.8:1 to 523:1. This differential behaviour of the COPL-pairs can be interpreted as follows. The six nouns with ratios below 10:1 are cases of relatively balanced overabundance according to the thresholds established in Thornton (2019a: 240–245). In contrast, those nouns whose ratios surpass this threshold (by far) instantiate unbalanced overabundance. Owing to the quantitative gap between the two groups of nouns, I conclude that – as in other cases discussed above – it is not possible to generalize sweepingly over COPLs in Romanian.

In Romanian, there are sets of three COPLs as shown in Table 10. Colloquial and infrequent options are highlighted by grey shading. The absolute token frequencies are generally relatively low. The order plural 1–3 reflects that given in Beyrer, Bochmann and Bronsert (1987: 82).

Table 10: Sets of three COPLs in Romanian.

singular	meaning	plural 1		plural 2		plural 3	
brîu	'belt, stretch of land'	brîie	35	brîuri	18	brîne	15
frîu	'reins'	frîie	14	frîne	69	frîuri	3
grîu	'grain'	grîne	73	grîiuri	0	grîie	2
				grîuri	1		

It strikes the eye that, except for the initial consonant, the segmental chains of the nouns in Table 10 are identical. Nevertheless, the options they prefer when it comes to pluralization are not the same. The COPLs of *brîu* 'belt, stretch of land' yield ratios which oscillate between 1.2:1 to 2.3:1. They are sufficiently low to justify speaking of relatively balanced overabundance. This classification is also licit for the ratio 4.9:1 for *frîne-frîie* (and 4.6:1 for *frîie-frîuri*). In the remaining cases, the ratios are considerably higher, namely 23.1:1 for *frîie-frîuri* and 36.5:1 for *grîne-grîie*. The plural *grîuri* is not mentioned in the reference grammar but attested once in the electronic corpus. The ternary sets corroborate the hypothesis that the COPLs do not behave homogeneously across the Romanian nouns.

In the electronic corpus, beside the survival of archaic plurals as e.g. SG *palat* 'palace' → PL *palate* (8,730 hits) ~ *palaturi*$_{ARCHAIC}$ (115 hits), there are certain neuters for which normative grammar has not yet decided upon the appropriate plural as e.g. SG *coltuc* 'bread crust' → PL *coltuce* (8 hits) ~ *coltucuri* (unattested) and SG *fetiş* 'fetish' → PL *fetişuri* (389) ~ *fetişe* (1).

Feminine and neuter nouns are especially prone to differentiate meanings in the plural. Table 11 hosts eight COPL-pairs of this kind.

In most of the cases, the semantic relation between the two plurals is metonymic. It is therefore legitimate to assume polysemy for the singular. The pair of sentential examples in (5) illustrates the differentiation of meanings in the plural for *ocol* 'detour; district'.[23] Note that there is gender alternation of neuter and feminine in the plural of this and other nouns.[24]

---

[23] In Beyrer, Bronsert and Bochmann (1987: 82), the meanings of the two plurals of *ocol* have been confused.

[24] For (5b), (6a), and (6b) some special characters of Romanian orthography have been corrected.

Table 11: Semantically conditioned overabundance in Romanian.

singular	plural 1	meaning	plural 2	meaning
bucată	bucăţi	'morsels'	bucate	'meals'
fantezie	fantezii	'fantasies'	fanteziuri	'silken cloth'
mîncare	mîncări	'dinners'	mîncăruri	'meals'
miros	mirosuri	'smells, odours'	miroase	'spices'
nivel	niveluri	'levels'	nivele	'floors'
ocol	ocoluri	'detours'	ocoale	'forestry districts'
oţel	oţel	'types of steel'	oţel	'gunlock'
plan	planuri	'plans'	plane	'levels'

(5) Romanian
  (a) detours [sfatubatranilor.ro] (Romanian Web 2016 (roTenTen16))
      Făceam **ocoluri** *lungi* *prin* *oraşul* *medieval.*
      make:IPF:1PL **detour:PL.N** long:PL.N through town:DEF.M medieval
      'We made long **detours** through the medieval city.'
  (b) districts [turistmania.ro] (Romanian Web 2016 (roTenTen16))
      Litera *"H"* *roşie* *marchează* *limita*
      letter:DEF.F H red:F mark.3SG boundary:DEF
      dintre *două* **ocoale** *silvice.*
      between two:F **district:PL.F** forestal:PL.F
      'The red letter H marks the boundary between two forestry districts.'

The semantic differentiation as assumed by the authors of the reference grammar is not always strictly obeyed by native speakers of Romanian as the examples in (6) suggest.

(6) Romanian [primariamiroslava.ro] (Romanian Web 2016 (roTenTen16))
  (a) *mirosuri*
      am *observat* *ca* *pe înserat* *se* *fac*
      AUX.1SG observe:PART that at dusk RFL make
      simţite **mirosuri** *grele* *de* *la* *fose*
      feel:PART:PL.F **odour:PL** heavy:PL.F of in ditch:PL.F
      '[] I have noticed that at dusk strong **odours** from the ditches become noticeable.'
  (b) *miroase*
      containarul *de* *gunoi* *din* *Gaureni* *este* *plin* *de* *gunoi*
      container:DEF of waste from Gaureni be.3SG full of waste

> şi **miroase** foarte urât.
> and **odour:PL** very ugly
> '[] the rubbish container from Gaureni is full of waste and very ugly **odours**.'

The two plurals *mirosuri* and *miroase* are used interchangeably with identical semantics referring to unpleasant olfactory impressions. Both utterances stem from the same website (https://www.primariamiroslava.ro/). On the one hand, this indiscriminate usage speaks in favour of the hypothesis that the formally different plurals belong to one and the same (polysemous) noun. On the other hand, if the one can replace the other in contexts like those in (6), it seems doubtful that this is a piece of evidence of semantically conditioned overabundance in the first place. There is a third possibility, namely irony – the speaker uses *miroase* in (6b) mockingly.

It is clear that Romanian COPLs call for further inspection. What can be said for now is that Romanian gives evidence of four different aspects of overabundance in the domain of plural marking. There is the stylistic factor with normative and colloquial variants of the plural. Ongoing language change marginalizes an older alternative. Normative grammar has not yet determined which of several options is to be declared official. Different readings of polysemous nouns are pluralized differently albeit with a certain margin of overlap with individual speakers.

4.5 Albanian

As in Romanian, there are COPLs in Albanian which have survived language reform. Buchholz and Fiedler (1987: 256) claim that

> [d]urch die Normung der letzten Jahre ist die Möglichkeit der Variierung von [Plural-]Bildungsweisen bei einem Lexem stark eingeschränkt worden. Waren noch in den sechziger Jahren bei Tausenden von Substantiven zwei, bei vielen Dutzenden zwischen drei und zehn verschiedenen Bildungsweisen anerkannt, so sind zwei Bildungsweisen (ohne wesentliche semantische Differenzierung) heute nur noch bei etwa 50 Subst[antiven] zugelassen.[25]

The number of COPLs thus has diminished drastically since the language reform of the early 1970s. Fiedler's (2007) revised and enlarged dissertation on the forma-

---

25 My translation: "owing to the recent standardization the options for pluralization of a lexeme have been severely restricted. In the 1960s, for many thousand nouns two, and for many dozens of nouns three to ten formations were still recognized, whereas today two formations (without semantic differences) are permitted only for some 50 nouns."

tion of the Albanian plural paints an enormously colourful picture of the variation across the Albanian speech community. For obvious reasons, I cannot do justice to the many regional and stylistic variants Fiedler mentions in his book-length study. In what follows, I take account only of those COPLs which are presently tolerated in the standard variety of Albanian.

In (7) I reproduce the list of "meaningless" COPLS as mentioned by Buchholz and Fiedler (1987: 256–257).

(7) Nouns with "meaningless" COPLs in Standard Albanian
SG *babagjysh* 'grandfather' → PL *babagjysha ~ babagjyshër*; SG *bajrak* 'banner' → PL *bajrakë ~ bajraqe*; SG *bakall* 'grocer' → PL *bakej ~ bakaj*; SG *bisht* 'tail' → PL *bishta ~ bishtra*; SG *buf* 'eagle owl' → PL *bufë ~ bufër*; SG *byrek* 'pie' → PL *byrekë ~ byreqe*; SG *çakall* 'jackal' → PL *çakej ~ çakaj*; SG *çam* 'Albanian from western Epirus' → PL *çamë ~ çamër*; SG *çap* 'step' → PL *çapa ~ çape*; SG *çorap* 'stocking' → PL *çorapë ~ çorape*; SG *dardhë* 'pear' → PL *dardha ~ dardhë*; SG *djep* 'cradle' → PL *djepa ~ djepe*; SG *drapër* 'sickle' → PL *drapinj ~ drapërinj*; SG *dyfek* 'rifle' → PL *dyfekë ~ dyfeqe*; SG *flámur ~ flamúr* 'flag' → PL *flámuj ~ flamúrë*; SG *gardh* 'fence' → PL *gardhe ~ gjerdhe*; SG *gisht* 'finger' → PL *gishta ~ gishtërinj*; SG *gjarpër* 'snake' → PL *gjarpërinj ~ gjarpinj*; SG *gjedh* 'cow' → PL *gjedhë ~ gjedhe*; SG *gjysh* 'grandfather' → PL *gjysha ~ gjyshër*; SG *hamall* 'porter' → PL *hamaj ~ hamej*; SG *hell* 'spit' → PL *heje ~ hej*; SG *hendek* 'ditch' → PL *hendekë ~ hendeqe*; SG *jastëk* 'cushion' → PL *jastëkë ~ jastëqe*; SG *kec* 'young buck' → PL *keca ~ kecër*; SG *krap* 'carp' → PL *krep ~ krapa*; SG *lab* 'inhabitant of Labëria' → PL *labë ~ lebër*; SG *lanet* 'devil' → PL *lanetë ~ lanetër*; SG *lepur* 'hare' → PL *lepuj ~ lepura*; SG *lojë* 'game' → PL *lojëra ~ lojna*; SG *mezé* 'snack' → PL *mezé ~ mezéra*; SG *mijë* 'thousand' → PL *mija ~ mijëra*; SG *mret* 'lime tree' → PL *mreta ~ mrete*; SG *muslluk* 'tap' → PL *musllukë ~ muslluqe*; SG *nip* 'nephew' → PL *nipa ~ nipër*; SG *ortek* 'avalanche' → PL *ortekë ~ orteqe*; SG *oxhak* 'stove' → PL *oxhakë ~ oxhaqe*; SG *poganik* 'birthday party' → PL *poganikë ~ poganiqe*; SG *princ* 'prince' → PL *princa ~ princër*; SG *prind* 'father' → PL *prindër ~ prindë*; SG *sanxhak* 'Sanjak' → PL *sanxhakë ~ sanxhaqe*; SG *sarëk* 'turban' → PL *sarëkë ~ sarëqe*; SG *sënduk* 'chest' → PL *sëndukë ~ sënduqe*; SG *sokak* 'alley' → PL *sokakë ~ sokaqe*; SG *stap* 'stick' → PL *stape ~ stapinj*; SG *stërgjysh* 'great-grandfather' → PL *stërgjysha ~ stërgjyshër*; SG *stërnip* 'great-grandchild' → PL *stërnipa ~ stërnipër*; SG *ulluk* 'gutter' → PL *ullukë ~ ulluqe*; SG *xham* 'glass' → PL *xhama ~ xhame*; SG *yll* 'star' → PL *yje ~ yj*; SG *zarf* 'envelope' → PL *zarfe ~ zarfa*.

The nouns in (7) do not constitute a semantically homogeneous class. The presence of numerous borrowings from Turkish is in line with the high number of Turkish loanwords in Albanian (Boretzky 1975).[26] The COPLs always come in pairs, i.e. more extended chains of COPLs are unattested. The order in which the COPLs are presented for the above nouns is that of Buchholz and Fiedler (1987). Ten nouns are taken randomly from the inventory in (7) to determine whether the COPLs differ in terms of their token frequencies according to the Albanian National Corpus (2011–2016 version) (consulted on 13 December, 2021). Figure 4 discloses the results for pairs of plural 1 + plural 2.

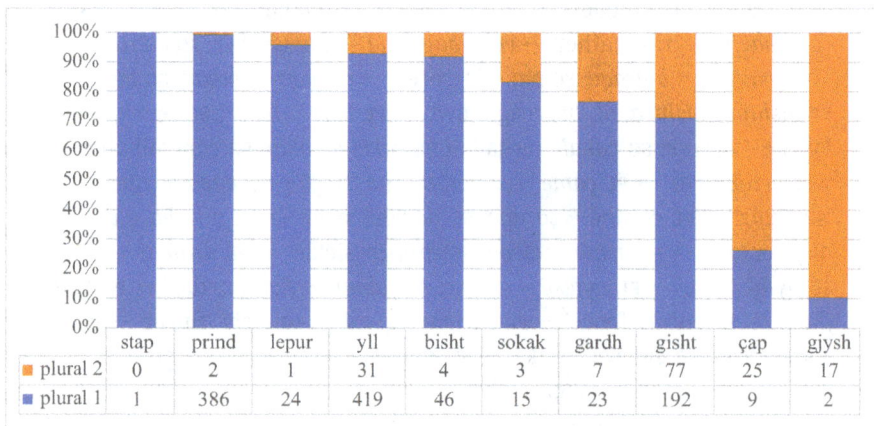

	stap	prind	lepur	yll	bisht	sokak	gardh	gisht	çap	gjysh
plural 2	0	2	1	31	4	3	7	77	25	17
plural 1	1	386	24	419	46	15	23	192	9	2

Figure 4: Token frequencies of ten selected Albanian COPLs.

The nouns yield different results for the quantitative relation between the COPLs. Figure 4 shows that the nouns can be ordered so that a continuum emerges. At one extreme, there is the generally infrequent noun *stap* which is attested only with one of the plurals. Towards the opposite pole, we find seven nouns which prefer plural 1 over plural 2 and two nouns which give preference to plural 2. It is remarkable that always one of the COPLs is responsible for the majority of the tokens. The relations thus tend to be unbalanced.

This interpretation receives support from another set of ten nouns in (7), namely those which display the voiceless velar plosive /k/ as stem-final segment. For these nouns, there is competition between plurals in *-ë* and *-qe* (with palatalization of /k/ > /c/ = <q>). These nouns do not form a semantically defined class but

---

26 The retention of the Turkish pluralizer *-ler/-lar* is reported for a small number of borrowed nouns such as SG *sheh* 'sheikh' → PL *shehlérë* (Buchholz and Fiedler 1987: 251). Dialectally, the Turkish pluralizer may also attach to Albanian stems such as SG *mbret* 'king' → PL *mbretër* (standard) ~ *mbretlerë* (regional) (Gardani 2012: 88).

share the phonological property of a final syllable rhyme with a vowel of different quality followed by a specified single coda consonant. Outside the domain of overabundance, the change of the stem-final voiceless velar plosive to the voiceless palatal plosive is a relatively frequent pattern under pluralization (Buchholz and Fiedler 1987: 258–264). On the basis of the dictionary dating back to the 1950s Buchholz and Fiedler (1987) assume that there are 30 nouns which reflect the pattern sɢ *armik* 'foe' → ᴘʟ *armiq*. About 100 nouns follow the pattern sɢ *bark* 'belly' → ᴘʟ *barqe*. More frequent are, however, /k/-final nouns which escape palatalization such as sɢ *mjek* 'physician' → ᴘʟ *mjek-ë*. Figure 5 features the frequency count for the *k*-final nouns with ᴄᴏᴘʟs in (7).

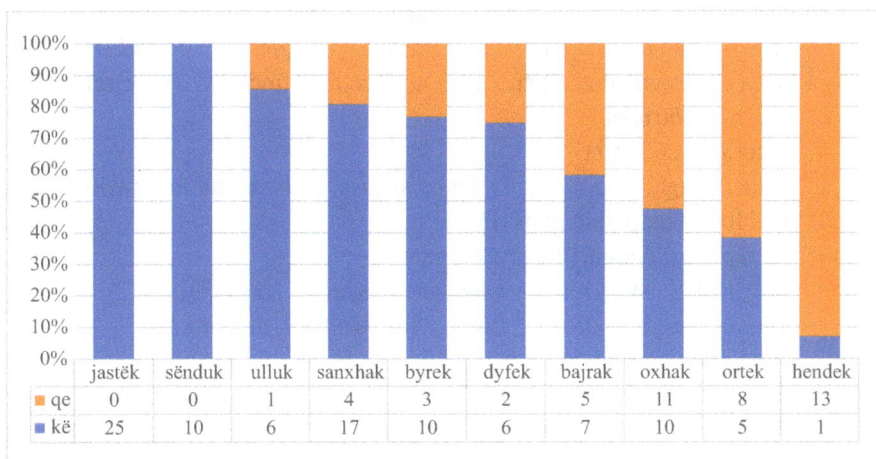

	jastëk	sënduk	ulluk	sanxhak	byrek	dyfek	bajrak	oxhak	ortek	hendek
▪ qe	0	0	1	4	3	2	5	11	8	13
▪ kë	25	10	6	17	10	6	7	10	5	1

Figure 5: Token frequencies of Albanian ᴄᴏᴘʟs (*k*-final nouns).

The picture which emerges from Figure 5 resembles that of Figure 4. The *k*-final nouns display individual preferences for one plural form over the other. Only with sɢ *oxhak* 'stove' → ᴘʟ *oxhakë* (48%) ~ *oxhaqe* (52%) are the shares of the competing plurals similar enough to pass as an uncontroversial case of balanced overabundance.

Buchholz and Fiedler (1987: 257) state that ᴄᴏᴘʟs are only infrequently employed to differentiate meanings. This statement notwithstanding, the turnout for "meaningful" ᴄᴏᴘʟs in the modern standard of Albanian can hardly be termed negligible. The descriptive grammarians postulate two systematic patterns:

(a) the plural suffix *-ër* is attached to nouns denoting animals to express metaphorical meanings as sɢ *hajvan* '(domestic) animal; blockhead' → ᴘʟ *hajvanë* ≠ ᴘʟ *hajvanër* 'blockheads' (= examples in (8)),

(b) for feminine nouns ending in stressed *-í/-é/-á/-ó*, there are ᴄᴏᴘʟ-pairs consisting of a pragmatically neutral zero plural (= syncretic with the singular) and a plural in *-ra* which highlights heterogeneity and often has deprecative connotations such as sg *çudí* 'miracle' → PL *çudí* 'miracles' ≠ PL *çudira* 'various miraculous things' (= examples in (9)).

The metaphorical use of the noun *ujk* 'wolf' in the sense of 'dangerous person' is formally distinguished from its normal reference to the animal species in the plural as shown in (8).

(8) Albanian (http://albanian.web-corpora.net/albanian_corpus/search)
  (a) zoonym [Lumi i madh]
      Solli   *edhe*   *shumë*   *gjarpërinj,*   **ujq**   *e*   *dhelpra*
      bring.AOR.3SG also   many   snake:PL   **wolf.PL**   and   fox:PL
      të   *mbytura,*   *që*   *i*   *kish*   *gjetur*   *strukur*
      ACC.PL   drowned:PL   REL   ACC   AUX.3SG.IPF   find:PART   take_refuge
      nëpër   *shtrofka*
      among   den:ACC.PL
      'S/he brought also many drowned snakes, **wolves** and foxes which s/he had found hidden in dens [].'
  (b) metaphor [Polonia, unë dhe bjondet]
      ju   *jeni*   *si*   *një*   *kope*   *me*   **ujqër**   *mendjesh*
      2PL   be.2PL   like   INDEF   flock   with   **wolf:PL**   thinking:ABL
      të   *pista.*
      ABL.INDEF.F   dirty:PL
      '[] you are like a pack of **wolves** with dirty minds.'

In (8a), *ujq* 'wolves' refers to the predators which took refuge in dens after a flood destroyed their natural habitat. In contrast, the wolves to which *ujqër* refers in (8b) are humans who behave like wolves in their social interactions.

The pair of sentences in (9) involves the COPLs of *idé* 'idea'.

(9) Albanian (http://albanian.web-corpora.net/albanian_corpus/search)
  (a) neutral plural [Agjencia Telegrafike Shqiptare]
      Kostandini   *u*   *martirizua*   *nga*   *forcat*
      Kostandini   PASS   martyrize:AOR.3SG.PASS   from   force:PL.DEF
      osmane   *me*   *prerje*   *të*   *kokës*   *për*
      Ottoman:PL   with   cutting   GEN.DEF head.GEN.DEF   for
      idetë   *e*   *tija*   *përparimtare*   *dhe*   *mëmëdhetare.*
      idea:DEF.PL GEN.DEF   POSS.3SG.M.PL   progressive:PL   and   patriotic:PL
      'Kostandini was martyrized by decapitation by the Ottoman forces for his progressive and patriotic **ideas**.'

(b) distributive [Koha.mk]

Ahmet Zogu	*ka*	*bërë*	*studime*	*serioze*	*në*	*Evropë,*
Ahmet Zugu	AUX.3SG	make.PART	study:PL	serious:PL	in	Europe

di	*disa*	*gjuhëra*	*të*	*huaja*	*dhe*	*është*
know.PRS.3SG	several	language:PL	ACC	foreign:PL	and	be.3SG

një	*njeri*	*me*	***idera***	*përparimtare.*
INDEF	man	with	**idea:PL**	progressive:PL

'Ahmet Zogu has studied seriously in Europe, he knows several foreign languages and is a man with progressive **ideas**.'

The COPLs are modified by the same postnominal adjectival attribute *përparimtare* 'progressive' in both sentences. In (9a), Kostandini's ideas which cost him his life are marked as definite plural by the suffixed *-të*. The fatal ideas are conceived of as a coherent ideological system. In contrast, Ahmet Zogu – president and king of interwar Albania – is characterized as a man with progressive ideas in (9b) for which the plural *idera* is chosen. The *ra*-plural seems to indicate that the politician had various ideas which, however, did not form a full-blown ideology or political program. Since the source from which (9b) has been taken is not critical of Ahmet Zogu, it cannot be concluded that the *ra*-plural has deprecative connotations in this particular case.

Neither the (a)-type nor the (b)-type of Albanian COPLs constitutes overdifferentiation. This is immediately clear for the metaphorical relations mentioned for (a). The *ra*-plural in (b) is reminiscent of the distributive whose exclusion from the domain of number categories has been argued already in Section 4.1. In addition, the *ra*-plural seems to be pragmatically loaded.

Beside these relatively systematic cases, there is a plethora of COPL-pairs which cannot be captured by either (a) or (b). Table 12 contains all examples provided by Buchholz and Fiedler (1987: 257). Their list is clearly marked as not exhaustive. Grey shading identifies cases of singular-plural syncretism. The singular forms in the leftmost column have been added on the basis of Rrahmani (1981).

Table 12: Semantically conditioned overabundance in Albanian.

singular	plural 1	meaning	plural 2	meaning
bar	*barëra*	'weeds'	*barna*	'medicines'
bostan	*bostanë*	'melons'	*bostane*	'bed of melons'
brinjë	*brinjë*	'ribs'	*brinja*	'slopes'
copë	*copë*	'pieces'	*copëra*	'fragments'
dru	*dru*	'sticks, timber'	*drurë*	'trees'
element	*elemente*	'chemical elements'	*elementë*	'criminal elements'

Table 12 (continued)

singular	plural 1	meaning	plural 2	meaning
farë	*fara*	'seeds'	*farë*	'types'
fund	*funde*	'ends'	*fundra*	'remains'
grykë	*grykë*	'throats'	*gryka*	'precipices'
gjak	*gjakra*	'bloodstains'	*gjaqe*	'vendettas'
gji	*gjinj*	'breasts'	*gjira*	'gulfs'
kokë	*kokë*	'heads of cattle'	*koka*	'heads'
lëkurë	*lëkurë*	'skins, hides'	*lëkura*	'barks'
milion	*milionë*	'X million'	*miliona*	'millions'
pikë	*pikë*	'points'	*pika*	'drops'
rob	*rob*	'family member'	*robër (lufte)*	'prisoner of war'
rreth	*rrethe*	'districts'	*rrathë*	'circles'
vit	*vjet*	'calendric years'	*vite*	'times'
zot	*zotërinj*	'masters'	*zota*	'gods'

The vast majority of the COPLs form binary sets. Buchholz and Fiedler (1987: 257) mention additional forms for two nouns, namely *farna ~ farëra* 'seeds' for *farë* 'seed' and *copa* 'pieces' for *copë* 'piece.' These additional examples invoke an interpretation as collectives. The range of meanings covered in Table 12 is such that neither the singulars nor the competing plurals define a conceptually coherent domain (Thornton 2010: 16–20). In (10), two sentences illustrate the different meanings of the COPLs of *bar* 'grass.'

(10)   Albanian (http://albanian.web-corpora.net/albanian_corpus/search)
    (a)   herbs [Gazeta Shqiptare]

          Emrat   *e*   *medikamenteve*   *ndryshojnë* *në* *varësi*
          name:PL GEN.DEF medicine:GEN   change:3PL in dependence
          të   *specialitetit*   *të*   *pavijonit,*   *por* *që* *në*
          GEN.PL speciality:GEN GEN.DEF pavilion:GEN.DEF but REL in
          total *janë*   *plot*   *235* **barna**
          total be:3PL exactly 235 **grass:PL**
          'The names of the drugs change according to the speciality of the pavilion so that there are exactly **235 drugs** in total.'

    (b)   narcotics [Koha.mk]

          Pakot   *me*   **barëra** *janë*   *nxjerrë*   *nga* *furgoni*
          package:PL with **grass:PL** be.3PL extract:PART from van
          dje   *në* *afërsi* *të*   *Tetovës*
          yesterday in vicinity GEN.DEF Tetovo:GEN
          'Packages with **(illegal) drugs** have been extracted from a van in the vicinity of Tetovo [].'

There is a semantic split. Legal drugs, herbs, and weeds can be referred to with the plural *barna* whereas narcotics and illegal drugs are often associated with the use of *barëra*. This semantic differentiation was probably not yet established when Buchholz and Fiedler (1987) published the reference grammar of Albanian. However, the semantic boundaries between the COPLS is often blurred. Similar to the Romanian case exemplified in (6) above, semantically supposedly distinct members of a COPL-pair can be used interchangeably as shown in (11).

(11)   Albanian (http://albanian.web-corpora.net/albanian_corpus/search)
   (a)   [Zëri]

Që	*nga*	*ajo*	*kohë,*	***gjakrat***	*janë*	*qetësuar*
REL	from	DEM.PRX.F	time	**blood:PL.DEF**	be:3PL	calm:PART

dhe	*frika*	*e*	*'islamizimit',*	*'orientializimit'*
and	fright:DEF	GEN	Islamism:GEN.DEF	Orientalism:GEN.DEF

e	*mani*	*të*	*tjera*	*janë*	*pakësuar.*
and	mania	GEN	other:PL	be:3PL	diminish:PART

   'Since this time the **vendettas** have calmed down and the fears of 'Islamism', 'Orientalism' and other manias have diminished.'

   (b)   [Panorama]

Këtu	*u*	*falen*	*36*	***gjaqe***
here	PASS	pardon:AOR.3PL.PASS	36	**blood:PL**

   'Here 36 **vendettas** were pardoned [].'

The noun *gjak* 'blood' is invested with COPLS. According to Table 12 the meanings of *gjakra* 'bloodstains' and *gjaqe* 'vendettas' are distinct. However, in (11a–b) both of the COPLS are used with the meaning 'vendettas.' It cannot be ruled out completely that the domains of some of the other semantically conditioned COPLS overlap in a similar way.

As to the COPLS in Albanian, the final word has not been spoken yet. The same holds for the Romanian case discussed in the foregoing section – and more or less also for all other languages mentioned in this study. What comes clearly to the fore is that both "meaningless" and "meaningful" COPLS are recurrent phenomena across a sizable number of languages. If many languages in different parts of the world give evidence of COPLS, this shared property calls for being studied thoroughly for as many languages as possible. Why a project of this kind makes sense is explained in the conclusions.

5 Conclusions

The cross-linguistic data presented in this study support the hypotheses put forward by Thornton (2010) on the basis of her Italian COPLs. Most of the facts speak against classifying COPLs as cases of overdifferentiation or defectivity. As to the former possibility, no instance of an additional number value could be identified although collectives and distributives seem to be involved in COPLs repeatedly, i.e. concepts which are related to number are represented in several COPL-pairs. Since only clear cases of polysemy have been taken account of, the possibility of defectivity has been precluded beforehand. The singular of the nouns displays all the meanings which are formally distinguished in the plural. It is impossible to define the classes of nouns which allow for COPLs on semantic grounds. Similarly, it is difficult to identify a particular conceptual domain for the meanings expressed by the COPLs. These findings need to be put to the test on a considerably larger empirical basis.

The same proviso holds for all further results of this study. We have seen that the unbalanced relation of the members of a set of COPLs is frequently connected to
- metaphor – (at least) one of the plurals is mainly used for the metaphorical meanings of a polysemous noun,
- style – (at least) one of the plurals is typical of formal or colloquial style,
- register – (at least) one of the plurals is used predominantly in the written or spoken register,
- genre – (at least) one of the plurals is used mostly in the discourse of a certain profession,
- change – (at least) one of the plurals is a relic of an earlier stage of the language,
- contact – (at least) one of the plurals has entered the system by way of borrowing.

The opposite effect on COPLs can be attributed to standardisation processes in the sense that many-to-one relations are often considered to be unnecessary structural complications of language by normative grammarians. Their prescriptive ideal may impel them to cancel alternative plurals from the norm. Thornton (2012b: 462–465) demonstrates, however, that overabundance may prove to resist normative regulations for a long time. Chances are that languages whose norm is devoid of COPLs allow for COPLs outside the standard.

Binary sets of COPLs are particularly numerous. However, more extended chains of COPLs are attested in several languages. The COPLs differ formally in various ways. Affixal pluralizers have been shown to compete with each other on one and the same stem. Introflection, internal modification, prosodic properties, and periphrasis are involved in the formation of COPLs whereas genuine suppletion of stems is attested only rarely. Fusional languages as well as agglutinating languages are affected by the phenomenon under inspection. Favourable conditions for the emergence of COPLS are the existence of inflectional classes and the preservation of relics of an erstwhile distinct number category.

Thornton (2019a: 254) characterizes the canonical overabundant cell of a paradigm as follows:

(a) "it is the only overabundant cell in a lexeme's paradigm"

This criterion is relevant for nouns which inflect for several categories. The Lower Sorbian (Section 4.2) nominative plural meets the expectations since it is the sole locus for COPLS to occur in this sample language.

(b) "no other lexemes of the relevant word class in the same language have overabundance in the same cell"

Mecayapan Nahuatl *huehuetquej* ~ *huehuetmej* 'elders' is the only case which conforms to this condition (Section 3.3).

(c) "it contains two (or more) forms that are in a 1:1 frequency ratio in a corpus"

This condition is not normally met by the COPLS reviewed in this study. Figures 1–5 (Maltese, Dutch, Romanian, Albanian) show that ratios of this kind are largely exceptional.

(d) "the usage of either one of the cell mates is not subject to any conditions (neither geo-socio-stylistic nor grammatical)"

Some of the COPL-pairs – such as the initial *sarar* ~ *soror* 'bundles' in (1a–b) from Maltese – seem to instantiate free variation. In many other cases it is hard to exclude that social, stylistic, or other conditions are at work.

In sum, the overwhelming majority of the COPLS discussed in this paper fail to meet all of the above criteria. If we discount Mecayapan Nahuatl *huehuetquej* ~ *huehuetmej* 'elders', none of the cases realizes the canonical overabundant cell. This failure notwithstanding, they attest to overabundance (with different degrees of canonicity). The definition of canonicity is such, however, that no specific relation to (high) frequency is presupposed (Corbett 2005).

The above analyses strongly suggest that the typologically-minded study of COPLS is still in its infancy, in a manner of speaking. The cross-linguistic recurrence of the phenomenon is such that it deserves to be paid more attention because knowing that COPLS show up in many places on the linguistic map of the world is not sufficient to declare the case closed. The frequent exploitation of COPLS for stylistic purposes, metaphor, collectives, and distributives, etc. renders the phenomenon especially interesting for linguists who are interested in number, morphological mismatches, semantics, and style. Thus, the inquiry into the nature of COPLS should be continued not only in typological perspective but also in-depth for individual languages.

Abbreviations

1/2/3	1st/2nd/3rd person
ABL	ablative
ACC	accusative
AOR	aorist
AUX	auxiliary
C	consonant
CMPR	comparative
COPL	co-plural
DAT	dative
DEF	definite
DEM	demonstrative
DIR	directional
ERG	ergative
F	feminine
FUT	future
GEN	genitive
INDEF	indefinite
IPF	imperfect
IRR	irrealis
LK	linker
M	masculine
N	neuter
NOM	nominative
NP	noun phrase
PART	participle
PASS	passive
PL	plural
POSS	possessive
PRX	proximate
REL	relative
RFL	reflexive
SG	singular
V	vowel

References

Acquaviva, Paolo. 2008. *Lexical Plurals. A Morphosemantic Approach*. Oxford: Oxford University Press.

Andersen, Torben. 2014. Number in Dinka. In Anne Storch & Gerrit J. Dimmendaal (eds.), *Number – Constructions and Semantics. Case Studies from Africa, Amazonia, India and Oceania*, 221–264. Amsterdam/Philadelphia: Benjamins.

Aquilina, Joseph. 1987. *Maltese-English Dictionary. Volume one: A–L*. Malta: Midsea.

Aquilina, Joseph. 1991. *Maltese-English Dictionary. Volume two: M–Z and addenda*. Malta: Midsea.

Aronson, Howard I. 1990. *Georgian: A Reading Grammar*. Columbus, OH: Slavica Publishers.

Benzing, Johannes. 1985. *Kalmückische Grammatik zum Nachschlagen*. Wiesbaden: Harrassowitz.

Beyrer, Arthur, Klaus Bochmann & Siegfried Bronsert. 1987. *Grammatik der rumänischen Sprache der Gegenwart*. Leipzig: Enzyklopädie.

Blevins, James P. 2008. Declension classes in Estonian. *Linguistica Uralica* 44(4). 241–267.

Boretzky, Norbert. 1975. *Der türkische Einfluss auf das Albanische*. 2 Teile. Wiesbaden: Harrassowitz.

Borg, Albert & Marie Azzopardi-Alexander. 1997. *Maltese*. London/New York: Routledge.

Bornemann, Eduard & Ernst Risch. 1978. *Griechische Grammatik*. Frankfurt am Main/Berlin/München: Diesterweg.

Buchholz, Oda & Wilfried Fiedler. 1987. *Albanische Grammatik*. Leipzig: Enzyklopädie.

Calder, George. 1980. *A Gaelic Grammar*. Glasgow: Gairm Publications.

Chung, Sandra. 2020. *Chamorro Grammar*. Santa Cruz, CA: University of California.

Corbett, Greville G. 2000. *Number*. Cambridge: Cambridge University Press.

Corbett, Greville G. 2005. The canonical approach in typology. In Zygmunt Frajzyngier, Adam Hodges & David S. Rood (eds.), *Linguistic Diversity and Language Theories*, 25–49. Amsterdam/Philadelphia: Benjamins.

Cunha, Celso & Lindley Cintra. 1989. *Nova gramática do português contemporâneo*. Lisboa: Sa da Costa.

Donaldson, Bruce C. 1993. *A Grammar of Afrikaans*. Berlin/New York: Mouton de Gruyter.

Eisinger, Marc. 1998. *Index lexical du Codex de Florence*. Thèse présentée pour le Doctorat en Linguistique de l'Université Paris 7 Denis Diderot. Tome I–IV.

Fähnrich, Heinz. 1986. *Kurze Grammatik der georgischen Sprache*. Leipzig: Enzyklopädie.

Fähnrich, Heinz. 1994. *Grammatik der altgeorgischen Sprache*. Hamburg: Buske.

Favereau, Francis. 1997. *Grammaire du breton contemporain*. Morlaix: Skol Vreizh.

Fiedler, Wilfried. 2007. *Die Pluralbildung im Albanischen*. Prishtinë: Akademia e Shkencave dhe e Arteve e Kosovës.

Gardani, Francesco. 2012. Plural across inflection and derivation, fusion and agglutination. In Lars Johanson & Martine Robbeets (eds.), *Copies versus Cognates in Bound Morphology*, 71–98. Leiden/Boston: Brill.

Geerts, Guido, Walter Haeseryn, Jaap J. de Rooij & Maarten C. van den Toorn. 1984. *Algemene Nederlandse Spraakkunst*. Groningen/Leuven: Wolters-Noordhoff.

Givón, Talmy. 1985. *Functionalism and Grammar*. Amsterdam/Philadelphia: Benjamins.

Grevisse, Maurice. 1964. *Le bon usage. Grammaire française avec des remarques sur la langue française d'aujourd'hui*. Gembloux & Paris: Duculot – Hatier.

Haiman, John. 2011. *Cambodian / Khmer*. Amsterdam/Philadelphia: Benjamins.

Helias, Per Jakez. 1986. *Dictionnaire breton*. Paris: Garnier.

Jacobs, Neil G. 2005. *Yiddish. A Linguistic Introduction*. Cambridge: Cambridge University Press.

Janaš, Pětr. 1976. *Niedersorbische Grammatik*. Bautzen: Domowina.

Jenny, Mathias & San San Hnin Tun. 2016. *Burmese. A Comprehensive Grammar*. London/New York: Routledge.

Khojayori, Nasrullo & Mikael Thompson. 2009. *Tajiki Reference Grammar for Beginners*. Washington, DC: Georgetown University Press.

Ladd, D. Robert, Bert Remijsen & Caguor Adong Manyang. 2009. On the distinction between regular and irregular inflectional morphology: evidence from Dinka. *Language* 85. 659–670.

Laskowski, Roman. 1972. *Polnische Grammatik*. Leipzig: Enzyklopädie.
Launey, Michel. 2011. *An Introduction to Classical Nahuatl*. Cambridge: Cambridge University Press.
Leger, Rudolf. 1994. *Eine Grammatik der Kwami-Sprache (Nordostnigeria)*. Köln: Köppe.
Lorenz, Manfred. 1982. *Lehrbuch des Pashto (Afghanisch)*. Leipzig: Enzyklopädie.
Mark, Colin. 2004. *The Gaelic-English Dictionary / Am Faclair Gàidhlig-Beurla*. London/New York: Routledge.
Matras, Yaron. 2009. *Language Contact*. Cambridge: Cambridge University Press.
Miestamo, Matti. 2008. Grammatical complexity in a cross-linguistic perspective. In Matti Miestamo, Kaius Sinnemäki & Fred Karlsson (eds.), *Language Complexity. Typology, Contact, Change*, 23–42. Amsterdam/Philadelphia: Benjamins.
Mifsud, Manwel. 1996. The collective in Maltese. *Rivista di Linguistica* 8(1). 29–52.
Nurmio, Silva. 2019. *Grammatical Number in Welsh: Diachrony and Typology*. Oxford: Wiley Blackwell.
Olko, Justyna, Robert Borges & John Sullivan. 2018. Convergence as the driving force of typological change in Nahuatl. *Language Typology and Universals (STUF)* 71(3). 467–507.
Portelli, Sergio & Sandro Caruana. 2018. Observing Eurolects: the case of Maltese. In Laura Mori (ed.), *Observing Eurolects. Corpus Analysis of Linguistic Variation in EU Law*, 268–293. Amsterdam/Philadelphia: Benjamins.
Rrahmani, Nazmi. 1981. *Fjalor i gjuhës së sotme shqipe*. Prishtinë: Rilindja.
Rzehak, Lutz. 2019. *Tadschikische Studiengrammatik*. Wiesbaden: Reichert.
Saade, Benjamin. 2018. The distribution of short and long pronouns in Maltese. *Language Typology and Universals (STUF)* 71(2). 199–220.
Schembri, Tamara. 2012. *The Broken Plural in Maltese: A Description*. Bochum: Brockmeyer.
Stolz, Thomas. 2001. Ordinalia – Linguistisches Neuland. Ein Typologenblick auf die Beziehung zwischen Kardinalia und Ordinalia und die Sonderstellung von EINS und ERSTER. In Birgit Igla & Thomas Stolz (eds.), *Was ich noch sagen wollte. . . A multilingual Festschrift for Norbert Boretzky on occasion of his 65th birthday*, 507–530. (Studia Typologica 2). Berlin: Akademie-Verlag.
Stolz, Thomas & Benjamin Saade. 2016. On short and long forms of personal pronouns in Maltese. In Gilbert Puech & Benjamin Saade (eds.), *Shifts and pattern in Maltese*, 199–268. (Studia Typologica 19). Berlin: De Gruyter Mouton.
Stolz, Thomas & Maja Robbers. 2016. Unorderly ordinals. On suppletion and related issues of ordinals in Europe and Mesoamerica. *Language Typology and Universals* (STUF) 69(4). 565–594.
Storch, Anne & Gerrit J. Dimmendaal. 2014. One size fits all? On the grammar and semantics of singularity and plurality. In Anne Storch & Gerrit J. Dimmendaal (eds.), *Number – Constructions and Semantics. Case Studies from Africa, Amazonia, India and Oceania*, 1–32. Amsterdam/Philadelphia: Benjamins.
Taimanglo, Roland L.G., Aline Yamashita & Maria A.T. Rivera (eds.). 1999. *Mandidok yan manfabulas na hemplon Guåhan*. Hagåtña/Guam: Government of Guam, Department of Education.
Thornton, Anna M. 2010. La non canonicità del tipo it. *braccio // braccia / bracci*: sovrabbondanza, diffetività o iperdifferenziazione. *Studi di grammatica italiana* 29/30. 419–477.
Thornton, Anna M. 2012a. Overabundance in Italian verb morphology and its interactions with other non-canonical phenomena. In Thomas Stolz, Hitomi Otsuka, Aina Urdze & Johan van der Auwera (eds.), *Irregularity in Morphology (and beyond)*, 251–270. (Studia Typologica 11). Berlin: Akademie.

Thornton, Anna M. 2012b. La sovrabbondanza nei paradigmi verbali dell'italiano contemporaneo. In Patricia Bianchi, Nicola De Blasi, Chiara de Caprio & Francesco Montuori (eds.), *La variazione nell'italiano e nella sua storia. Varietà e varianti linguistiche e testuali, Atti dell'XI Congresso SILFI,* 445–456. Firenze: Cesati.

Thornton, Anna M. 2013. Compagni di cella in una gabbia dorata: sull'uso di vo vs. vado nell'italiano contemporaneo. In Emili Casanova & Cesáreo Calvo Rigua (eds.), *Actes du XXVIe Congrès International de Linguistique et de Philologie Romanes (València, 6–11 de septembre de 2010),* 447–458. Berlin: De Gruyter.

Thornton, Anna M. 2019a. Overabundance: a canonical typology. In Franz Rainer, Francesco Gardani, Wolfgang U. Dressler & Hans Christian Luschützky (eds.), *Competition in inflection and word-formation,* 223–258. Berlin: Springer.

Thornton, Anna M. 2019b. Overabundance in morphology. *Oxford Research Encyclopedia of Linguistics* (April 2019). [DOI: 10.1093/acrefore/9780199384655.013.554]

Trepos, Pierre. 1982. *Le pluriel breton.* Brest: Brud Nevez.

Tschenkéli, Kita. 1958. *Einführung in die georgische Sprache. Band I: Theoretischer Teil.* Zürich: Amirani.

Van den Toorn, Maarten C. 1977. *Nederlandse grammatica.* Groningen: Wolters-Noordhoff.

Viitso, Tiit-Rein. 2003. Structure of the Estonian Language. In Mati Erelt (ed.), *Estonian Language,* 9–129. Tallinn: Linguistica Uralica.

Vulanović, Relja & Oliver Ruff. 2018. Measuring the degree of violation of the one-meaning–one-form principle. In Lu Wang, Reinhard Köhler & Arjuna Tuzzi (eds.), *Structure, Function and Process in Texts,* 67–77. Lüdenscheid: RAM.

Weil, Gotthold. 1917. *Grammatik der osmanisch-türkischen Sprache.* Berlin: Reimer.

Wolff, H. Ekkehard. 1993. *Referenzgrammatik des Hausa.* Münster: LIT.

Wolgemuth, Carl. 1981. *Gramática nahuatl del municipio de Mecayapan, Veracruz.* México/DF: Instituto Lingüístico de Verano.

Yousef, Saeed. 2018. *Persian. A Comprehensive Grammar.* London/New York: Routledge.

Electronic corpora

Albanian National Corpus. 2011–2016 version, http://albanian.web-corpora.net/
Dutch Web 2014. (=nlTenTen 14), https://www.sketchengine.eu/nltenten-dutch-corpus/
Estonian Web 2019. (=etTenTen 19), https://www.sketchengine.eu/ettenten-estonian-corpus/
Korpus Malti 3.0, https://mlrs.research.um.edu.mt/CQPweb/
National Corpus of Polish, http://nkjp.pl/poliqarp/nkjp300/query/
Romanian Web 2016. (=roTenTen16), https://www.sketchengine.eu/corpora-and-languages/romanian-text-corpora/

Silva Nurmio

5 Towards a typology of singulatives

Definition and overview of markers

Abstract: The singulative is a category that has never been mapped out crosslinguistically. This article begins to do that, starting with a definition and a disambiguation of terminology related to grammatical methods of individuation and unitization. The singulative is argued to be one strategy to denote individuals and units, and it occurs as part of many different number systems. I present a typological overview of markers, both those which are straightforward to analyse as singulatives and more ambiguous examples. This allows us to test and define the boundaries of this category and its overlap with others (e.g. the diminutive). I also suggest some diachronic developments which further illustrate the relationship between the singulative and other categories.

Keywords: singulative, grammatical number, morphology, linguistic typology, typology of number, diminutive

1 Introduction

This paper has two aims: to define what singulatives are and are not, and to give a typological overview of singulative markers. The focus on markers allows us to apply the definition to examples and to define the boundaries of this category, i.e. what does and does not count as singulative. It brings to light possible diachronic patterns in the development of singulatives and the marking of number and individuation more generally. It also demonstrates how this category can overlap with the diminutive and the implications that this has for the definition and typological search for singulatives.

The paper begins with examples and a definition in 1.1. Section 2 compares the singulative to other categories used for individuation, unitization and packaging and argues for their disambiguation. Section 3 focuses on defining and finding singulatives: I examine the treatment of singulatives in previous literature, introduce the new database, and present the terminology used in the paper. Section 4

Acknowledgements: I would like to thank Rahel T. Dires, Matilda Carbo, Don Killian, Thomas Stolz and Iwan Wyn Rees for providing feedback to various parts of this article. All remaining errors are mine alone.

https://doi.org/10.1515/9783110986600-005

examines different ways of marking the singulative cross-linguistically. Section 5 concludes the paper.

1.1 Definition and terminology

(1) presents three nouns from Welsh (Indo-European). (1a) shows the pattern taken by the majority of Welsh nouns, with an unmarked singular and a marked plural (plurative [PLT] in my terminology, see below); this is the same pattern we find with most English nouns, e.g. *cat~cats*. A relatively small group of nouns, illustrated by (1b), have the opposite pattern, with an unmarked plural and a marked singulative (abbreviated SGT). (1c) is an example of a mass noun, some of which can also have a singulative denoting a unit out of the mass.

(1)   a.   *cath* 'cat', PLT *cath-od* 'cats'
       b.   *moch* 'pigs', SGT *moch-yn* 'a pig'
       c.   *gwellt* 'grass, straw', SGT *gwellt-yn* 'a blade of grass, a (single) straw'

More examples of singulatives are given in (2) and (3) from a typological range, based on the Singulative database (Nurmio, Dires & Carbo forth.). The base forms of singulatives (such as *dimín* in (2a)) are not glossed for their grammatical number status here for simplicity; this varies, as does terminology, and the exact status is often difficult to determine from grammars (discussed further below). The examples show nouns with singulatives, but note that not all nouns in these languages mark number in this way:

(2)   a.   *dimín*, SGT *dímina* 'cloud' (Khamtanga, Afro-Asiatic, Appleyard 1987: 254)
       b.   *ūl*, SGT *úldis* 'drop of water' (Ket, Yenisean, Vajda 2022: 329)
       c.   *kuk*, SGT *kukte* 'piece of charcoal' (Bari, Nilotic, Spagnolo 1933: 45)
       d.   *neĝo*, SGT *neĝero* 'snowflake' (Esperanto, Artificial language, Kellerman 2005: 134, 431)

Here are a few example sentences with singulatives. (3a-b) illustrate units of aggregate substances (pluralized in (3b)) while the singulative (3c) denotes an individual, similar to (1b) above. (3c) also shows possible agreement effects of singulatives: in Arabic, singulative nouns are feminine and trigger feminine singular agreement on the verb. This is not generalisable to all languages with singulatives; I will not go further into agreement in this paper. Likewise, there is much to be said about the semantic basis of singulative marking, but that is another topic for another paper.

(3) a. *Jisa   ajweba-mū kwla*
         pick.up rice-SGT   all
         'Pick up all the rice kernels.' (Baule, Atlantic-Congo, adapted from Timyan
         1976: 163)
     b. *Last-h   jïh   lopme-tjelmie-h   hispieh*
         leaf-PLT and   snow-SGT-PLT   whirl.3PL.PRES
         'Leaves and snowflakes are flying around.' (South Saami, Uralic, adapted
         from Ylikoski 2022)
     c. *Nemel-a   kaḥl-a   daxl-et   l-el   kujina*
         ant-F.SGT black-F.SG   entered.3-F.SG   to-the   kitchen
         'A black ant entered the kitchen.' (Arabic, Afro-Asiatic, Dali & Mathieu
         2021: 278)

I define the singulative as follows:

(4) A singulative is a derived noun form which is formed by adding a marker to a
     non-unit-denoting base and which denotes 'a/one X' or 'a/one unit of X'

Many singulative markers have other meanings in addition to individuation ('a/
one X') and unitization ('a/one unit of X'), such as having a diminutive function
with some bases, as discussed below. Consider (4) in the light of the Welsh example
in (1). In (1b), the base is a morphologically unmarked plural form: its syntactic
behaviour matches that of marked plurals such as *cath-od* 'cats' (Nurmio 2019: 59;
Plein 2018: 287–310) but, unlike *cath-od*, *moch* bears no plural marker. In (1c) the
base is a mass noun, i.e., it cannot be directly modified by numerals in its basic form
(Nurmio 2019: 145).[1] The singulative is formed by adding a suffix which has two
forms in Welsh: *-en* (feminine) and *-yn* (masculine). A suffix is the most common
singulative marker (see section 4) but the definition includes any kind of morphological marking. The term "singulative" seems to go back to Zeuss' *Grammatica
Celtica* (1853: 299–301). In this passage we also find the term "collectives" (*formae
collectivae*) applied to bases like *moch* 'pigs' which continues to cause terminological confusion in Brittonic linguistics and beyond (see Nurmio 2017: 63–65 for a

---

1 A note about frequency is needed here. Singulatives formed from mass nouns such as (1c) are
rare in Present-day Welsh. For many mass nouns, e.g. *tywod* 'sand', the singulative appears to not
be used, and the more common way of denoting units is with the measure phrase *gronyn o X* 'a
grain of X', e.g. *gronyn o dywod* 'a grain of sand' (with a regular mutation on the noun). The main
dictionary for Welsh, *Geiriadur Prifysgol Cymru* [GPC] (ed. Thomas et al.), continues to list singulatives for such nouns (e.g. *tywodyn* for 'a grain of sand') but this may not reflect present-day usage.

discussion).[2] The function of singulatives ranges from singular number ('a pig' as opposed to 'pigs') to unitization ('a blade of grass'); singulatives may take part in inflectional number distinctions or they can be a derivational strategy.[3] Unitization can be further divided into natural minimal units ('*grain* of sand'—I will use this example from now on since *grain* is a more prototypical minimal unit noun than e.g. *blade* in English) and packaged units ('chocolate *bar*'), as discussed below.

The definition in (4) states that singulatives exist in opposition to non-unit-denoting bases. In our database (Nurmio, Dires & Carbo, forth.), these include e.g. unmarked plurals and mass nouns as seen in (1). Another common type is "general number", defined by Corbett (2000: 10) as forms whereby "the meaning of the noun can be expressed without reference to number". This is illustrated here from the Ethiopian language Sidaama (Afro-Asiatic) (Kawachi 2007: 345):

(5) BASE   SGT   PLT
     sina 'branch(es)'   *sin-čo* 'a branch'   *sin-na* 'branches'

Such forms may be number-neutral, as the glossing above suggests, but not always: Mous (2021: 523–524) observes that some are, in fact, associated (more) with a plural or a singular meaning. I strongly agree with Mous on the need to study the pragmatics of these forms further: it is rare to find examples of e.g. their agreement behaviour in grammars, not to mention variation in usage. Since it is not clear to what extent these forms have truly general meaning, I follow Mous (2021) and Dires (forth.) in calling them "base (forms)". Terminology abounds for these: terms gathered from the literature on Cushitic languages alone include the following (see Dires, forth.): base (form), basic, collective, general, generic, mass, singular, trans-numeral, unit, unmarked. This shows the need to clearly distinguish morphology from semantics. Such forms may indeed refer to collections or mass substances, but I use "base" specifically to refer to the form (unmarked, in opposition to at least a singulative or plurative form).[4] The term "collective" occurs particularly often in relation to singulatives across languages, and it remains a topic for further research

---

2 In Nurmio (2017, 2019) I use the term "morphological collective" which is arguably a useful descriptive term for Welsh since these nouns are not simply a plural allomorph but have characteristics that warrant classifying them as a separate category. However, I use "unmarked plural" as a comparative concept in this paper to facilitate cross-linguistic comparison.

3 In translations of singulatives into English, I propose using the article ('*a pig*') to distinguish singulatives clearly from "generic" nouns ('pig' referring to a type of animal, with no reference to number).

4 In more detailed treatments of Cushitic-style systems, it may be wise to disambiguate base (= general term in morphology) and base (= a member of noun paradigms such as that in (5)) by e.g. capitalising the latter.

whether it emerges as a *morphological* category in some languages or whether it is best reserved as a semantic term only.

The bases described so far are all forms that are words on their own, such as *sina* 'branch(es)' in (5). There is another type of paradigm, the so-called replacive or replacement pattern. The term was coined by Dimmendaal (2000a, 2021) for "Nilo-Saharan" languages.[5] (6a) shows an example from Mursi (Surmic) (Worku 2020: 256). A similar pattern also occurs in Brittonic; (6b) is from Welsh (Nurmio 2019: 134). Mursi also has the patterns unmarked singular/marked plurative and unmarked plural/marked singulative; in this, it also parallels Welsh.

(6) a. SGT(?)   PLT   b. SGT(?)   PLT
        ɲàb-ì 'ear' *ɲàw-à* 'ears'   *blod-yn* 'flower' *blod-au* 'flowers'

So, crucially, the replacement pattern typically seems to exist alongside the plural/singulative pattern, and I am not aware of a language with only the replacement pattern. Dimmendaal (2000a: 216) calls this type of number marking the "tripartite pattern" since there are three kinds of marking: singulative, plurative (Dimmendaal uses "plural") and replacive/replacement. The question is, do we still define the member of the pair with singular meaning as singulative in (6) (hence the question marks)? Here, the base is not a noun that occurs on its own synchronically, but rather a stem (*ɲàb-* and *blod-*). I suggest treating these as singulatives, albeit less prototypical singulatives than those covered so far, for the main reason that the same markers are used in both the unmarked plural/singulative pattern and the replacement pattern in these languages. Secondly, it may be that for some nouns this pattern derives diachronically from the unmarked plural/singulative pattern. There are likely to be several reasons why nouns end up in the replacement pattern (see Nurmio 2019: 134–136 for a discussion of Welsh) and this is another topic requiring more research.

---

5 There is disagreement whether several language families with shared features, including Surmic, belong to a larger macrofamily called Nilo-Saharan, see Güldemann (2018: 235–309) and Dimmendaal (2000a: 216-217). Glottolog (Hammarström & Forkel 2021) does not use Nilo-Saharan in its classification scheme. In this paper, the classifications in brackets follow Glottolog, with the understanding that some classifications are not fixed. In fact, the status of tripartite number as an areal vs. genealogical feature is an important part of the debate (Güldemann 2018: 255–256).

2 Singulative vs. other strategies for individuation, unitization and packaging

There are other morphological unitization devices which are not singulatives in my definition but are sometimes collated with them. In my definition (see (4) above), *marker* refers to bound markers and not lexemes. This excludes words such as *grain* in *grain of sand*, or *jyvä* 'grain' in Finnish (Uralic) *hiekanjyvä* [sand.GEN grain] 'a grain of sand' since these are not bound markers, but lexemes attested on their own. However, it is possible for such lexemes to grammaticalize into singulative markers (see 4.1 below). The terminology for the *grain*-type words varies, e.g. "unit counter" (Greenberg 1972: 12; Allan 1977: 293), "minimal unit word/noun" (Goddard 2010: 146) and "individuating or partitive (sense)" (Jurafsky 1996: 555). I follow Goddard in calling them minimal unit nouns, since these nouns refer to the perceivable minimal units of aggregates.[6] Jackendoff (1991, see also Corbett 2000: 79–80) proposes a classification of different types of nouns based on two semantic variables: boundedness and internal structure. All mass nouns are characterized by lacking the feature +bounded (since one can add or remove sand and still have *sand*), but they can be divided into two further groups as regards +/-internal structure: aggregates (e.g. sand) have it, substances (e.g. water) do not. In terms of semantics, then, minimal unit nouns are formed from mass nouns with the feature +internal structure. A related type of marking is seen in *a head of cattle* or Breton (Indo-European) *penn-moc'h* [head-pigs] 'a pig' with *penn* 'head' (see 4.2 below). This usage of 'head' is similar to that of minimal unit nouns, with the distinction that the referent is animate. Since one head of cattle is the smallest natural unit that makes up cattle, I propose calling *head* a minimal unit noun in the *head of cattle* construction (a subtype for animate nouns, to be more specific).

Another type of construction found with mass nouns and comparable to the singulative (see e.g. Haspelmath & Moravcsik 2019) is "measure words/nouns". Such nouns are used with numerals and other quantifiers to measure out the referent, e.g. *one/a **cup** of coffee, three **kilos** of sand, one/a **piece** of cake*. Such nouns can be used to measure out both countable objects (*three kilos of apples*) and mass substances. This is not the case with minimal unit nouns which only apply to mass nouns with internal structure (**a grain of apples* is not possible). Greenberg (1972: 13) also observes that the *grain*-type unit nouns are almost completely restricted to a usage with 'a/one' and not higher numerals, while measure nouns do not have this restriction. Like minimal unit nouns, measure nouns are lexemes attested

---

6 Perceivable in the sense of human conceptualisation (cf. Corbett 2000: 79); we are not concerned with minimal units in the sense that physicists would be.

on their own, and are therefore to be kept apart from singulatives, despite some overlap in function. It should also be noted that, like classifiers (see below), unit nouns and measure nouns have *meaning* beyond just 'a unit of X' and 'a measure of X', since they are lexical nouns. Even *grain*, the meaning of which is somewhat bleached in *grain of sand* since this does not denote the grain of a plant, still carries meaning beyond unitization, i.e. it is "something very small" and "something hard" (Goddard 2010: 146–147).

The term "(numeral) classifier" is also sometimes found applied to nouns that I just defined as measure nouns (e.g. Lehrer 1986; Nurmio 2019: 137 uses the tentative term "classifier-like construction"). In this paper, I follow Allan (1977) in restricting the term "classifier" to classifier languages, such as Thai (Thai-Kadai) or Mandarin Chinese (Sino-Tibetan). In such languages, the majority of nouns require an additional element—a classifier—to precede them before they can be modified by e.g. numerals. Such languages have two types of constructions, one of which is often called "mensural classifier", shown in (7b) (Aikhenvald 2000: 114–115). These add a unit of counting/measuring to the noun modified (e.g. 'pile of X') and are the type that overlaps most with measure nouns, since they apply to both semantically count and mass nouns. Note that some authors, e.g. Crisma et al. (2011: 285) who have a typological/comparative approach, call mensural classifiers "measure expressions", avoiding the term classifier (as reflected in the glossing of (7b) below).

"Sortal classifiers" are usually treated as classifiers proper: they modify nouns that are semantically count (+bounded) and denote "one" or, more accurately, "times one" (Greenberg 1972: 10). In this, they resemble singulatives. However, in addition to unitizing, classifiers "have meaning" (Allan 1977: 285), most common meanings having to do with animacy and shape (Crisma et al. 2011; Aikhenvald 2000; Croft 1994). (7a) shows the Mandarin noun *shū* 'book' with its sortal classifier which has a shape reference (Crisma et al. 2011: 285).

(7) a. *yī*   *běn*   *shū*   b. *yī*   *duī* *shū*
     one CL^{VOLUME} book   one pile book
     'one/a book'   'one/a pile of books'

Since the classifier is required for the noun to be modified by numerals, it is a unitizing device. Classifiers are an interesting parallel to singulatives since they share the characteristic of denoting individuals or units. The difference is the fact that classifiers add meaning beyond just individuation/unitization. There are also examples such as Yucatec Maya (Mayan) *há'as* 'banana' which takes on a different meaning with a different classifier ('banana *fruit*' vs. 'banana *tree*' vs. 'banana *leaf*' etc.) (Lucy 1992: 74; see Seifart forth. for a similar example with noun class markers in Miraña (Boran)). Such examples can indeed involve unitization ('banana' (non-number specified) > 'a banana' (single fruit)) but there is also an added meaning (a fruit as opposed to e.g. tree). Also, as noted by Greenberg (1972: 7), classifiers are often nouns themselves, not bound markers, which is another reason not to equate them with singulatives.

There is yet another type of unitization process, seen in the English examples *beer* (mass) vs. *a beer, two beers* (count). This operation is called recategorization by Corbett (2000: 81); other terms include coercion (from mass to count) (e.g. Dali & Mathieu 2021: 276). In such examples, something that is more commonly treated as mass is treated as count and this is shown in the noun phrase syntax. This operation is also called the Universal Packager (Pelletier 1975; Jackendoff 1991). Recategorization shares some features with singulatives formed from mass bases, but it is not a singulative formation, since the recategorization from mass to count is not brought about by any marker. Rather, one can assume a recategorized count noun *beer* (= a conventional unit of beer), which can then be used with numerals, etc. Also, as noted by Dali & Mathieu (2021: 276), nouns resulting from recategorization can often refer to both kinds ('one type of beer') and units ('one bottle/pint of beer'), while singulatives appear to have only the latter function.

I have defined the singulative as a form that is morphologically derived from its base. As also noted by Acquaviva (2016: 1172), this leaves out pairs such as *cow~cattle* and *person~people* from this definition, although, again, they are interesting in the wider typology of how groups and individuals thereof are denoted.

In addition to denoting units from substances with natural minimal parts (such as grains of sand or blades of grass), singulatives have a related function, namely creating packaged meanings from substances without natural minimal parts. Consider:

(8) a. *caws* 'cheese', SGT *cos-yn* 'a cheese' (with a regular vowel change) (Welsh, Indo-European, Nurmio 2019: 72)[7]

b. *šokolad* 'chocolate', SGT *šokolad-k-a* 'a chocolate bar' (Russian, Indo-European, Kagan & Nurmio forth.)

c. *bier* 'beer', SGT *bier-tje* 'a beer' (Dutch, Indo-European, Acquaviva 2016: 1172)

Acquaviva (2016: 1172) discusses this function of the singulative separately, but this distinction is not made in all discussions and often examples of the two are

---

7 Note that *cosyn* is not normally used for 'a piece of cheese' (contra Cuzzolin 1998: 129; Acquaviva 2016: 1180) but only for a complete packaged cheese (such as one might buy in a supermarket). For the former, *darn o gaws* [piece of cheese] 'a piece of cheese' is more likely. However, these impressions need to be tested with more speakers, since it is possible that there is, in fact, more variation.

listed together. Greenberg (1972: 12–13) does note this difference, namely 'piece of meat'-type (which he calls "unit counters") and 'grain of sand'-type ("particulates" in his terminology). Many instances involve markers with overlapping diminutive and singulative functions. The singulative is here similar in function to the Universal Packager mentioned above (Pelletier 1975; Jackendoff 1991), which packages the referent into a conventionalised form, such as a (packaged block of) cheese or a chocolate bar. Such examples of singulatives appear to be rarer than those denoting natural minimal units (as seen in 1c): for instance, Nurmio (2019: 77–84) lists 66 singulative forms in Welsh, with *caws~cosyn* as the only example of the packaging type. Examples such as those in (8) are also more lexicalised than singulatives which denote natural minimal units; for example, *šokoladka* is a chocolate bar, not e.g. a chocolate praline (another conventional packaged form of chocolate). In Kagan & Nurmio (forth.) we discuss more Russian examples (see also Acquaviva 2016: 1178), noting that there may be more nuances to this function. Some packaged meanings are closer to natural units since they are perceived as the smallest possible unit of which the derivative is true: for example, Russian *marmeladka* 'fruit jelly candy' (from *marmelad* 'marmalade') is one piece of candy, and it cannot be divided further and still be called *marmeladka*. This is a topic that requires more research.

The Brittonic languages have a small group of derivatives formed with the same marker as the singulative, and with an idiosyncratic function. For example:

(9) a. *gwiniz* 'wheat' — *gwiniz-enn* 'a wheat field' (Breton; Acquaviva 2008: 245; Trépos 1980: 67)
   b. *tywod* 'sand' — *tywod-en* 'sandbank, sand-dune' (Welsh; GPC: s.v. *tywoden1*)

These derivatives denote not minimal units of the referent, as expected, but features of landscape related to the referent. Welsh *tywoden* has only two attestations in GPC, both from the eighteenth century and both from dictionaries. Google search yields no results and native speakers consulted were not familiar with this form.[8] It is not a productive formation. The examples in Trépos (1980: 67; 1982: 269) are more interesting, since his work is informed by contemporary usage, including dialect variation. Strikingly, he reports that forms like *gwiniz-enn* have both the unit meaning 'stalk of wheat' and the "extent of" (*étendue de*) meaning 'wheat field' (Trépos 1980: 67) shown in (9a). I have not come across such a function of singu-

---

8 There is, however, a dialect form *dwnan* 'sand dune', with the dialect variant -*an* of the singulative/diminutive ending -*en* (Nurmio 2019: 119), which likewise refers to an extent of a particular landscape type (GPC: s.v. *twyn*; Iwan Wyn Rees pc.). This is based on a count noun *twyn* 'hill; dune', rather than a mass noun, but should nevertheless be considered in further studies of this phenomenon.

lative markers in other languages and it is not productive in Welsh either (and probably also not in Breton). I therefore refrain from suggesting that this is another function of the singulative cross-linguistically but wish to draw attention to these examples should similar ones be found in other languages.

This brief overview is not meant to be exhaustive. Rather, I have included this discussion to show the kind of morphological and semantic strategies which are similar to the singulative and sometimes confused with it, and to argue for a terminology that keeps different formations apart. This is particularly important since the singulative is a construction that is lacking in English, and it may be this very fact that has led to terminological confusion, as we might try to describe this phenomenon using existing terms such as "measure word". Table 1 summarises this discussion, setting the singulative apart from these other types of unitizing constructions.

Table 1: Summary of the categories for individuation, unitization and packaging discussed in this section, markers in bold.

Term	Denotes	Example
singulative	an individual a minimal unit a packaged unit	Welsh *moch-**yn*** 'a pig' Welsh *gwellt-**yn*** 'a blade of grass, a (single) straw' Welsh *cos-**yn*** 'a cheese'
unit noun	a minimal unit	English *a **grain** of sand, a **head** of cattle*
measure noun	a measured/ packaged unit	English *a **cup** of coffee*
mensural classifier	a packaged unit	Mandarin Chinese *yī **duī** shū* 'one/a pile of books'
sortal classifier	individual	Mandarin Chinese *yī **běn** shū* 'one/a book'
recategorization	a packaged unit	English *a beer*

The table brings into focus the similarity of singulatives with unit nouns, and, to a lesser extent, measure nouns as well as classifiers (both mensural and sortal). The difference is that singulatives are formed with bound markers and do not have meaning beyond unitization, i.e. they do not denote characteristics such as the shape of the referent, as classifiers do. In addition to overlap in function, there is a possible diachronic continuum between different unitization devices; I demonstrate in section 4 below that it is possible for singulative markers to develop from unit nouns through grammaticalization. In theory, measure nouns and classifiers could also be possible sources. Crisma et al. (2011: 286) and Yue (2016: 115) discuss the Mandarin Chinese classifier *ge* which shows signs of becoming a default classifier in several dialects for nouns previously (and prescriptively) associated with other classifiers. Yue (2016: 115) notes that this shift has, in fact, happened in a dialect of Northern Chinese and shows signs of being completed in another, related one. In her grammar of Wutun, defined by Sandman (2016: 2) as a Northwest Mandarin variety with strong contact influence especially from Tibetan (it has also been previously described as a "mixed language"), *ge* has also become a default marker used with all nouns (Sandman 2016: 63–68 and pc.). As such, it may in fact qualify as a singulative: it is added to non-unit denoting bases to denote an individual and, being a default marker, no longer has further meaning such as shape or size of the referent. This, too, remains a topic for further research.

3 Finding singulatives

3.1 Singulative in previous literature

My definition in (4) largely aligns with Acquaviva (2016: 1171–1172) who states that "if a unit-denoting noun is morphologically derived from a more basic non-unit-denoting noun, the derived noun, its grammatical category, the individualizing marker, or the individualizing derivation, are often called *singulative*" (his italics). In my definition, there is more emphasis on derivation by a *marker* (to the exclusion of *grain of sand*-type derivatives). Secondly, my definition aims to reflect the two kinds of meanings seen in (1b–c): (1c) is indeed unit-denoting, but with count nouns like (1b), a single pig is perhaps better described as an individual than a unit.

The definition in (4) proposes the term singulative as a comparative concept (Haspelmath 2010), i.e. a term that should be applicable to any form that fulfils these criteria, despite language-specific descriptions often using different terms. It is common to find singulatives described as "singular" and in instances where the singulative overlaps with diminutive, different authors can call the same form diminutive or singulative (see Kagan & Nurmio, forth., for a discussion on Russian). Terminology matters, for several reasons. First, in order to study a phenomenon cross-linguistically, we must be able to compare like with like. Secondly, terminology matters to how we understand grammatical phenomena in individual languages. In Nurmio (2017, 2019) I review how Welsh singulatives are described in the literature, noting that calling them singular has led many scholars to try and lump these forms together with morphologically unmarked singulars, obscuring the fact that the number system of Welsh in fact has the interesting splits shown above in (1). Finally, as a result of the first two issues, our overall understanding of grammatical number remains incomplete as regards this phenomenon. Typological discussions tend to note that overtly marking the singular number value is rarer than overtly marking the plural, i.e. that singulatives are a relatively rare phenomenon (Greenberg 1966; Haspelmath & Karjus 2017: 1214–1215, 1217). They are clearly rarer than morphologically marked plurals (pluratives in my terminology), but they may not be as rare as has been thought, given that there has never been a comprehensive data collection. In fact, singulatives are part of the number systems of one of the biggest language families in the world (by language count and by speaker numbers, see Ethnologue), namely Afro-Asiatic. An appreciation of these issues will add to our understanding of the diversity and complexity of grammatical number by shifting our view away from a rather Indo-European and Eurocentric view that takes a singular-plurative (*cat–cats*) paradigm as the norm (cf. Storch & Dimmendaal 2014: 1–2). This is seen, for example, in grammatical descriptions that state that a plural is formed by deleting a marker from the singular, i.e. making the singular the starting point of the paradigm even though it is the morphologically marked form (see e.g. Appleyard's (1987: 254) description of the Khamtanga example in (2a), Williams (1980: 9, 13) for Welsh, and Arensen (1982: 40) for Murle (Surmic)).

As already noted, the singulative is an understudied category. While it is an established term for talking about especially the Brittonic branch of Celtic and about Afro-Asiatic languages, there is relatively little cross-linguistic work on it. Some existing discussions are Dali & Mathieu (2021), Grimm (2018), and Acquaviva (2016). In all of these, the typological sample is small; Dali & Mathieu (2021) look at Semitic, Celtic and Nilotic. Haspelmath & Karjus (2017: 1220) mention six languages with singulatives, and the generalisation that they are found "in substantial numbers" only in certain families—Celtic (Indo-European); varieties of Arabic, and Cushitic (Afro-Asiatic); and some other languages of north-eastern Africa. Grimm (2018) looks at four languages. Acquaviva (2016) is the most comprehensive treatment with 19 individual languages mentioned, as well as three families (Nilo-Saharan [see fn. 5], Cushitic and Chadic) mentioned as having singulatives. A full typological picture is lacking. This gap means that all hypotheses about the origin of singulative marking are awaiting to be tested. On a more practical level, it also means that singulatives go unrecorded and under-described in grammatical descriptions since grammar writers may not recognise them or look for them by e.g. checking whether a diminutive marker can be added to a mass noun to form a singulative.

3.2 Data collection and terminology

The data in section 4 below comes from the Singulative database (Nurmio, Dires & Carbo, forth.). It is the first dataset focused on this category, recording data on morphology and semantics. In this paper, I focus solely on markers. The scope is not limited to languages where the singulative is part of the inflectional number system, but I also discuss derivational types (such as those in (8)), especially singulatives formed from mass nouns.

Since the singulative is an understudied topic, it is perhaps for this reason that the term itself has not been firmly established. Formations that are singulatives in my definition go by various names in the literature[9], making typological data collection more challenging; this is especially true of non-productive and derivational singulative formations such as those formed from mass nouns with markers that more commonly mark the diminutive. My terminology for number marking here largely follows Haspelmath & Karjus (2017), Treis (2014) and Junglas & Güldemann (forth.) in separating morphological form from meaning and using "plurative" and "singulative" to denote morphologically marked forms, as opposed to unmarked singular and plural.[10]

Table 2 summarises my terminology, focusing on morphology rather than meaning. Since singular and plural (ditto dual, paucal etc.) are also semantic terms, the qualifier "unmarked" can be added to specify that the reference is to the *form* of the noun:

Table 2: Terminology used in this article.

Morphological forms (marking)	Example
(unmarked) singular	English *pig*
(unmarked) plural	Welsh *moch* 'pigs'
plurative	Sidaama *sin-na* 'branches'
singulative	Sidaama *sin-čo* 'a branch'
base (form)	Sidaama *sina* 'branch(es)'

Note that the terms "marked" and "unmarked" refer here to the synchronic state: a synchronically unmarked form is one which has no markers that can be deleted. Diachronically speaking, however, a form may bear a marker/markers. This is the case with the Welsh unmarked plural *llygod* 'mice' which historically has a plural suffix *-od*, but the old singular *llyg* has been overtaken by the singulative *llygod-en* in the basic meaning 'mouse' (Nurmio 2017: 64); the same goes for Breton *logod* 'mice' (Trépos 1980: 71). *Llyg* survives in the meaning 'shrew', including compounds such

---

9 These include singulative, unit noun, nomen unitatis, singular, among others.

10 As already noted in Corbett (2000: 17), if used logically and consistently, "plurative" should be used to refer to any plural form in opposition to a more basic singular, and as such this would include e.g. the English *s*-plural. This is unlikely to catch on given the established use of the term "plural", but it is a useful and necessary term when discussing systems which show variation in whether the singular or plural number value is morphologically marked. In English-type systems where unmarked singular/marked plural is the overwhelming majority pattern, it may be less necessary to make such a distinction.

as *llyg y dŵr* 'water shrew' (see termau.org). Breton examples also include *stered* 'stars' (unmarked plural), old plurative of now-lost Middle Breton *ster* (Jouitteau & Rezac 2016). Cushitic languages also have examples of base forms which are old singulatives, e.g. Alaaba-K'abeena *'arriččut(a)* 'sun' (Crass 2005: 64; Dires, forth.) can be analysed as historically containing the singulative suffix *-iččut(a)*. Separation of form and meaning has already been pointed out as crucial when studying singulatives, and to that I would add a call to keep synchrony and diachrony separate as well.

The Singulative database (Nurmio, Dires & Carbo, forth.) is currently at 258 different singulative markers (see Table 3) in 120 different languages and some collection work is still ongoing. The database also currently covers languages which, we hypothesize, may have singulatives, but this is yet to be confirmed due to information on number/unitization being incomplete or lacking altogether.[11] As these numbers show, one language can have more than one singulative marker; an example of a language with a large number of markers is Iraqw (Afro-Asiatic) at 21 singulative suffixes (Dires forth.). Since singulatives are not equally represented across language families, this is not a database built by sampling. Instead, we aim to include all attested examples of singulatives both in living languages and also those spoken in the past (including extinct languages and earlier language stages, e.g. Old Irish (Indo-European)). Data is collected from existing literature and often by checking sister languages of those that are recorded to have singulatives. Some data is also collected directly from language experts. As already noted, singulatives built from mass nouns can sometimes be found by looking at descriptions of diminutives.

4 Singulative markers

In this section, I list all types of singulative markers found in the database. The prevalence of each marking type is discussed and some preliminary discussion about the diachrony of markers is included when it has a bearing on the synchronic analysis. Table 3 shows different markers in the database and the number of languages that have it. The list is not exhaustive (see above) but is meant to indicate trends as observed in the data collection so far.

---

11 The numbers here should also not be taken as absolute for two further reasons: first, the line between what constitutes dialects of one language and separate languages is fluid in many instances. Secondly, it is not always easy to tell whether something is an allomorph of another marker or a separate marker, especially when working with short and/or old grammar sketches.

Table 3: Types of singulative markers per number of languages in Nurmio, Dires & Carbo (forth.); numbers as of 23 February 2023.

Marker	Nr of languages
suffix	109
prefix	7
circumfix	1
stem alternation	3

4.1 Suffixation

For nominal plural, suffixation is the most common marking strategy cross-linguistically (Dryer 2013). The same applies to singulative markers: 91% of the languages in the current database mark the singulative this way. For examples, see (1), (2) and (3) above.

Suffixation can take different forms. In the majority of instances, including examples (1)–(3) above, we can speak of "singulative suffixes" (sometimes with other overlapping functions such as the diminutive). However, there are also examples of inverse markers: they change the number value of the basic, unmarked form (Corbett 2000: 159; the term "inverse" goes back to Wonderly, Gibson & Kirk 1954). That is, the same marker can be used for different number values. This is illustrated in (10) with Dagaare (Gur) (Grimm 2021: 448, see also Grimm 2012, 2018), with the inverse marker *-rí*:

(10) a. *tié*   *tìì-**rí***   b. *nyágá*   *nyág-**rí***
        tree.BASIC tree.INVERSE   root.BASIC root.INVERSE
        'a tree'   'trees'   'roots'   'a root'

In (10a), the unmarked form of the noun is singular in meaning and *-rí* changes it to plural meaning. (10b) is the opposite pattern: the basic, unmarked form has a plural meaning and the form with *-rí* a singular one. *Nyág-rí* 'a root' conforms to my definition of singulative although *-rí* is not a singulative marker. Inverse number is also found in Nilotic, Austronesian and Kiowa-Tanoan languages (Corbett 2000: 156–165; see also Sutton 2010 for Kiowa-Tanoan); and in Atlantic-Congo (in Bidyogo, see Di Garbo & Agbetsoamedo 2018: 191). This kind of marking is, therefore, attested in a typologically diverse range of languages.

Singulative suffixes may have grammaticalized from minimal-unit nouns ('*grain of sand*' type) (cf. below for prefixes with a similar history). One such transparent example is the noun 'eye' being used to denote a unit of a mass substance in a group of Uralic languages and also the Yeniseian language Ket which has been in contact with the Uralic language Selkup (Ylikoski 2021). In some languages the marker is clearly analysable as the lexeme 'eye' (e.g. Khanty and Selkup, see Däbritz 2021: 96, 103), in which case it would not qualify as a singulative marker but rather a minimal unit noun. This seems also the best analysis for Hungarian -*szem* from *szem* 'eye' (as noted by Haspelmath & Moravcsik 2019). In others, this lexeme has reduced phonologically, suggesting grammaticalization towards a singulative marker proper. This is visible in Ket: in (11a) the suffix still appears analysable as *dēs* 'eye', while in (11b), it has developed to -*dis* (cf. also (2b) above). -*s* in (11c) is suggested by Helimski (2016) to derive from *dēs*, and if this analysis is correct, this would make it a grammaticalized singulative marker.

(11) a. *eːl* 'berries' — *eːl-des* 'a (single) berry' (Helimski 2016: 158)
      b. *hǝnaŋ* 'sand' — *hǝnaŋ-dis* 'a grain of sand' (Däbritz 2021: 113)
      c. *aʔq* 'wood, trees, forest'[12] — *ōk-s* 'a tree' (Helimski 2016: 159)

Ket and Selkup both have another singulative marker, Ket -*lamt* and Selkup -*laka,* -*laga*, going back to a noun 'piece' (Däbritz 2021: 103, 113), suggesting a measure noun function in origin. While the diachronic developments suggested here require more research, I highlight this example in order to draw out possible diachronic links between singulative markers and minimal unit nouns as well as measure nouns.

4.2 Prefixation

Singulative marking by prefixation is less common. As noted above, nominal plurality is most commonly expressed by suffixation, too, so this picture is not surprising. Prefixation as a pluralisation strategy is most common in Africa, specifically within Niger-Congo and especially Bantu languages (Dryer 2013). This generalisation also extends to singulatives, if we choose to include noun class markers which mark individuals and units under the definition of singulatives. Consider the following example from the Bantu language Nata (Atlantic-Congo) (Laine 2023: 62):

---

12 Helimski (2016: 160) notes that perhaps the substance meaning 'wood' should be the main translation.

(12) a. *oβu-ɾɛɛsa*   b. *oɾu-ɾɛɛsa*
        CL14-beard   CL11-beard
        'beard'   'a (single) beard hair'

In Bantu languages, noun class and number information are jointly conveyed by a noun class prefix (Contini-Morava 1999: 4). Singular and plural meaning pairs are, therefore, both morphologically marked. Some constitute regular singular-plural meaning pairs. Some, however, have more of a "lexical-derivational" type relationship in Crisma et al.'s (2011: 260–261) terms: this is found with class 11 which may be used to derive units from mass nouns, as seen in (12). To decide whether certain noun-class markers have a singulative function, one would need to be able to demonstrate that (12b) is derived from (12a). While (12a) is not morphologically unmarked (it bears a class 14 prefix), Bantu experts seem to suggest that this form is a more basic member of the paradigm. Gunnink (2018: 124), writing about Fwe, notes that a shift to class 11 expresses "a singular entity of something that usually does not occur by itself", implying that this form is semantically marked (and presumably less frequent in usage than its base). For Swahili, Acquaviva (2016: 1173, referring to Contini-Morava 1999) argues that noun class reassignment can have a singulative function. I leave open the question whether forms like (12b) are best described as singulatives or not; to do so may require amending the definition of singulative to include not just morphological but also semantic markedness.

We also find a singulative prefix, or one developing into such, in Breton. In addition to the singulative suffix *-enn*, cognate with Welsh *-en*, Breton has another way of denoting individuals from groups and aggregate substances, namely prefixing with the following nouns: *penn* 'head', *loen* 'animal' and *pezh* 'piece, bit' (Jouitteau 2009–2022: *les singulatifs, ar pezh*; Trépos 1982: 236; Nurmio 2019: 137; Acquaviva 2008: 244, 258).

(13) a. *moc'h* 'pigs' – *penn-moch, pe-moch* 'a pig'
        b. *kezeg* 'horses' – *loen-kezeg* 'a horse'
        c. *dilhad* 'clothes, clothing' – *pezh dilhad* 'a piece of clothing'

Loen and *pezh* are best analysed similarly to English *head (of cattle)* and *piece (of clothing)*; i.e. they are not singulatives according to definition (4) but rather unit nouns, since they are lexemes rather than bound markers. Welsh has a similar construction with *llwdn* (also spelled *llwdwn*) 'young of an animal; animal' (GPC: s.v. *llwdn*). GPC gives examples such as *llwdn ceffyl* 'young horse' where the reference is to young animals, but compare e.g. *llwdn dafad* 'sheep; wether' where there is no reference to age.[13] This is confirmed by a Google search, although the examples come from older sources and it is unclear how common this usage still is; this warrants further research.

Breton *penn* seen (13a) is, however, of more interest regarding singulatives. The form with *moc'h* 'pigs' varies between dialects from *penn-moc'h* to reduced *pe-moc'h* (also *pi-moc'h*). In this latter form, *penn* could be said to have reduced and grammaticalized to a singulative prefix *pe-*. This only applies to some Breton dialects and, to the best of my knowledge, only happens with *moc'h* 'pigs' (confirmed by Mélanie Jouitteau, pc.). The dialect atlas by Le Roux (1924–1953, s.v. *porc, porcs*) shows the reduced form to be more widespread than its non-reduced counterpart. *Penn-* is also used to derive one member of paired body parts denoted by Breton duals (called "minor duals" in Nurmio 2019): e.g. *daoulin* 'knees (of one person)' – *penn-daoulin* 'a knee'. I am not aware of examples where *penn-* phonologically reduces in such constructions.

In the Goidelic branch of Celtic, there is a somewhat similar usage with a prefix. These are formations with the noun *leath* 'half; side', used with natural pairs to emphasise that reference is to one member of the pair, e.g. Irish *leath-chluas* [half-ear] 'one ear'. There are typological parallels from various families (see Nurmio 2019: 38–39). In the Irish example, *leath* 'half; side' functions more like a prefix than a lexical noun, since its meaning has been bleached to 'one'; *leath-chluas* does not mean 'half an ear' or 'side of an ear'. This would qualify *leath-* as a singulative marker. However, synchronically the base is a singular noun *cluas* 'ear' and not a non-unit-denoting noun which would be required for it to be classified as a singulative under definition (4). Irish nouns for body parts are complex also in the non-singular meaning: the pair meaning is often expressed not by the plural but a minor dual construction 'two' + noun (in a form reflecting the otherwise lost dual), e.g. *dhá chluais* [two ear.DUAL] '(one person's) ears' (Nurmio 2019: 35–38). Therefore, while there is a regular number paradigm *cluas* 'ear', PLT *cluasa*, there is a parallel pattern, specifically with reference to the pair and one member of the pair: *dhá chluais* '(one person's) ears'/ *leath-chluas* 'one ear (of the pair)'. The latter are semantically in a paradigm-like relationship of "pair/one member of the pair". However, we lack usage data, and it is not clear whether speakers who use one of these also use the other, making it difficult to assess to what extent these are truly a paradigm. I leave the Irish 'half'-forms tentatively out of the definition for singulatives, but they are clearly a related and interesting category.

---

13 *Llwdn* on its own has been restricted to the sense 'wether' in northern Welsh (Fynes-Clinton 1913: 353). I am grateful to Iwan Wyn Rees for drawing my attention to the *llwdn*-construction.

4.3 Circumfixation

Circumfixation refers to marking by both a prefix and a suffix. Only one example occurs in my sample, in the Berber branch of Afro-Asiatic. This branch is variably presented as one language ("Berber") or a number of dialects/languages (see Kossmann 2020 for details); I use the term dialect here for convenience. Example (14) comes from Nafusi (also 'Nefusi') (Beguinot 1942: 32).

(14)   *azemmûr* 'olive(s)', SGT *t-azemmûr-t* 'an olive'

As is common in Afro-Asiatic, the marker used in the singulative function in (14) is also the marker for feminine gender. Some exceptions are listed in Kossmann (2013b: 218). Feminine/singulative suffixes with *-t* are common to the Afro-Asiatic branch, while the circumfix marking with a prefixed *ta-* appears to be unique to Berber to the best of my knowledge.[14] As for other Berber dialects, Kossmann (2013a: 20, 44; 2013b: 216) notes that the system seen in (14) is found in some, but not all of them (see also Di Garbo & Agbetsoamedo 2018: 203). Most Berber nouns have a singular/plurative opposition and only a subset (centred around terms for vegetables and fruits) have the system seen in (14), referred to in the literature as the collective/unity system (Kossmann 2013: 217). It has been suggested that this is an innovation in Berber which arose under influence from Arabic (while realizing the singulative form with Berber morphology) (Kossmann 2013b: 283).

4.4 Stem alternations

There are many instances of the addition of a singulative affix triggering regular stem alternations, e.g. Welsh *adar* 'birds', *ader-yn* 'a bird' with raising of the root vowel. However, since this alternation is triggered by the singulative suffix *-yn*, I only count the suffix as the singulative marker proper.

There are also instances where the singular value is distinguished from the plural value by stem alternation only, i.e. there is no affix. Such examples are seen in the Nilotic languages Dinka (Andersen 1990, 2014) and Nuer (Baerman & Monich 2021). In (15) I use the glosses "nominative singular/plural" used in the source literature since it is not currently clear to me whether these forms could be interpreted as singulative or plurative.

---

14 I'm grateful to Stanly Oomen for advice on Berber.

(15) a. Dinka
        NOM PL   NOM SG
        bḛ̀l   *bẹel*
        'sorghum (plural), canes'   'sorghum (singular), cane'
    b. Nuer
        NOM PL   NOM SG
        bĕɛl   *bẹ̈ẹl*
        'dura (type of sorghum)'   'dura (type of sorghum)'[15]

Singular and plural are distinguished by changes to one or more features of the stem. In Dinka, these are vowel quality, vowel length, voice quality of the vowel (breathy vs. creaky), tone and the final consonant (Andersen 2014: 223). No particular combination signals singular or plural number: although there are some recurring singular-plural patterns, this does not amount to a predictable system. The number of a noun can be reliably detected only by looking at agreement (Andersen 2014: 224, 230). So even though the paradigms in (15) might lead one to interpret the lengthening of the vowel and change to breathy voice as a singulative marker, this is not the case since neither change denotes singularity.

Dinka does not have suffixes but marks inflectional distinctions by changes to the stem only (Ladd et al. 2009: 660). Nuer, on the other hand, has case/number suffixes in addition to stem alternation patterns. Nuer has three cases (nominative, genitive, locative) and two numbers (singular, plural) (Baerman & Monich 2021). There are two suffixes, one singular and one plural, which are used to mark some case and number forms, while the other forms remain unsuffixed. The NOM.SG. form is always unsuffixed. In addition, nouns have different stem alternations according to case and number which are similar to those found in Dinka (vowel quality and length, and tone). Baerman & Monich (2021: 266–268) note that, similarly to Dinka, the inflectional system of Nuer reflects an earlier system with what they call basic singulars (where the plural [= plurative in my terminology] is formed from the singular) and basic plurals (where the singular [= singulative] is formed from the plural). This distinction is reflected in the fact that old basic plural forms never have a NOM.PL. suffix in Nuer. Crucially, they note that the basic singular/basic plural contrast does not appear to be productive in the language.

A comparison with sister languages suggests that diachronically Dinka and Nuer had the tripartite system typical for "Nilo-Saharan" languages (see above) where some nouns overtly mark the plural value, others the singular value, and

---

15 I follow the authors' glosses here, which differ slightly between the two languages.

some mark both with suffixation (see Dimmendaal 2000a and discussion above). Dinka has lost the number-marking suffixes altogether, while in Nuer suffixes only mark some case/number distinctions while stem alternations do the rest. Another Western Nilotic language Päri is included in (16) to show the cognate noun which has an unmarked plural and a marked singulative.

(16) Päri (Andersen 1990: 18)

PL   SGT
bɛ́ɛl   *bèel-ó*
'sorghum (plural), canes'   'sorghum (singular), cane'

Diachronically speaking, then, Dinka *bɛ́el* and Nuer *bɛ́ɛɛl* are ex-singulatives, and for this reason some authors analyse Dinka as having an underlying tripartite system. However, Andersen (2014: 244–252) argues that in the synchronic stage the singular may be becoming the default basic member of the number paradigm. This is evidenced in particular by borrowed and derived nouns, e.g. SG *ǎnɡ̣anɡ̣aṯ*, PLT *ǎnɡ̣anɡ̣aaṯ* 'pineapple' (< Arabic *ananaas*), where the plurative is formed from the singular by lengthening the vowel of the last syllable (Andersen 2014: 245; see also Ladd et al. 2009: 666–668). Baerman & Monich (2021: 265) similarly observe that recent borrowings to Nuer inflect through suffixation alone, with the NOM.SG. as the unmarked form.

Synchronically, the pattern in (15) could be interpreted as a type of inverse number marking: singular and plural meaning are distinguished from one another by changes in the root while none of these changes is exclusively associated with either number value. This system differs from the Dagaare-type inverse system discussed above in that there is a range of possible inverse markers rather than just one. For Dinka, Andersen (2014: 226) argues that this means that the singular cannot be predicted from the plural, or vice versa. This is supported by Ladd et al. (2009: 666, 668) who observe some patterns of "subregularities" but conclude that they do not amount to the choice of number-marking being predictable. For Nuer, Baerman & Monich (2021: 264) demonstrate some predictability. Another analysis might be to regard Dinka and Nuer as having a replacement system where both singular and plural values are morphologically marked. This is what Andersen (2014: 244) suggests tentatively for Dinka. I leave open the exact analysis of these systems; psycholinguistic work might shed more light on the question of predictability of one form from another and the question whether the form with singular meaning is becoming the base for the plural one. In any case, I conclude that synchronically (15) cannot be interpreted as a tripartite systems with singulatives, although they are relevant to understanding the history of such marking.

The Nilotic examples show that singulative markers may be deleted by sound changes. Another instructive example of this is Baale-Olam (Surmic) with the singulative marker *-c* which is not realized unless it appears before another suffix, i.e. it is not present in all environments (Dimmendaal 2000b: 194–195). Compare *sŏ* (SG)/ *sɔ̌* (PL) 'foot, leg' with no suffix following, but *só-j-a-naandí* 'my leg' (cf. *sɔ̌-a-ɠaandí* 'my legs') with a pronominal linker *-a-* and with the singulative realized here as *-j-*.

5 Conclusion

This paper proposed a definition of singulatives, followed by a discussion of other common grammatical strategies for individuation and unitization that share some features with singulatives but which I argued should be kept apart from them. I have presented an overview of different types of singulative markers based on a typological data collection. These are mostly suffixes, but we also find examples (some of them tentative) of prefixes, stem alternation patterns, and one circumfix. Some singulatives occur as part of the grammatical number system of a language, but singulatives can also be derivational; most commonly they derive units of mass nouns. The Welsh examples in (1b) and (1c) illustrate these functions in a language where both patterns are present. Especially in the latter function, examples may not be labelled as singulative in the literature, and descriptive terminology for singulatives varies greatly in general.

This overview of patterns is meant as neither exhaustive nor the final word on what kind of patterns fall under the term "singulative". Instead, it is an exploration of what kinds of patterns fall under my definition and what are left outside of it. Some patterns are more straightforward to analyse than others. I have argued that some individuating/unitizing patterns belong under other labels while others were shown to count as singulatives diachronically but not synchronically. I have left open the question of how some formations should be classified for now; much work remains to be done on individual languages and typologically.

While the diachronic development of singulative marking deserves its own paper, I have demonstrated some possible pathways in this overview. One is the grammaticalization of unit nouns (***grain*** *(of sand)* type) into singulative markers. I also tentatively suggest that it may be possible for a classifier system to reduce to just one classifier whose function is simply unitization, which may warrant its classification as a singulative. This, too, requires further research.

Arriving at a crosslinguistic definition and a preliminary picture of how widespread singulatives are has implications for future research. Firstly, it can aid with the identification of singulatives, which can otherwise get placed under labels such as "singular" or "diminutive" in language descriptions. Secondly, the overlap between the singulative and other categories is very common in my data, raising questions about the connection between these categories. Here I have only discussed the overlap with the diminutive; other features or categories include gender and definiteness. The data collection also indicates that markers that form singulatives also have a variety of derivational functions, and it remains to be studied to what extent we can find crosslinguistic patterns in this. Finally, singulatives are a promising case study for the diachrony of number marking and unitization.

Abbreviations

CL	classifier; noun class (for Bantu)
F	feminine
GPC	Thomas et al. (eds), *Geiriadur Prifysgol Cymru.*
NOM	nominative
PL	plural
PLT	plurative
SG	singular
SGT	singulative

References

Acquaviva, Paolo. 2016. Singulatives. In Peter O. Müller, Ingeborg Ohnheiser, Susan Olsen & Franz Rainer (eds.), *HSK Word-Formation. An international handbook of the languages of Europe*, 1171–1183. Berlin: De Gruyter Mouton.

Aikhenvald, Alexandra Y. 2000. *Classifiers: a typology of noun categorization devices.* Oxford/New York: Oxford University Press.

Aikhenvald, Alexandra Y. 2003. *A Grammar of Tariana, from Northwest Amazonia.* Cambridge: Cambridge University Press.

Allan, Keith. 1977. Classifiers. *Language* 53(2). 285–311.

Andersen, Torben. 1990. Vowel length in Western Nilotic languages. *Acta Linguistica Hafniensia* 22(1). 5–26.

Andersen, Torben. 2014. Number in Dinka. In Anne Storch & Gerrit J. Dimmendaal (eds.), *Number – Constructions and semantics: Case studies from Africa, Amazonia, India and Oceania*, 221–264. Amsterdam: John Benjamins.

Appleyard, David L. 1987. A grammatical sketch of Khamtanga—I. *Bulletin of the School of Oriental and African Studies* 50(2). 241–266.

Arensen, Jon. 1982. *Murle grammar.* Juba: College of Education, University of Juba, and Summer Institute of Linguistics, and Institute of Regional Languages.

Baerman, Matthew & Irina Monich. 2021. Paradigmatic saturation in Nuer. *Language* 97(3). 257–275.

Basset, André. 1952. *La Langue Berbère.* London: Oxford University Press.
Beguinot, Francesco. 1942. *Il berbero Nefûsi di Fassâto: grammatica, testi raccolti dalla viva voce, vocabolarietti.* Pubblicazioni dell'Istituto per l'Oriente. 2nd edn. Roma: Istituto per l'Oriente.
Contini-Morava, Ellen. 1999. Noun class and number in Swahili. In Ellen Contini-Morava & Yishai Tobin (eds.), *Between grammar and the lexicon*, 3–29. Amsterdam/Philadelphia: Benjamins.
Corbett, Greville G. 2000. *Number.* Cambridge: Cambridge University Press.
Crass, Joachim. 2005. *Das K'abeena: deskriptive Grammatik einer hochlandostkuschitischen Sprache.* Köln: Rüdiger Köppe Verlag.
Crisma, Paola, Lutz Marten & Rint Sybesma. 2011. The point of Bantu, Chinese and Romance nominal classification. *Rivista di Linguistica* 23(2). 251–299.
Croft, William. 1994. Semantic universals in classifier systems. *WORD* 45(2). 145–171.
Cuzzolin, Pierluigi. 1998. Sull'origine del singolativo in celtico, con particolare riferimento al medio gallese. *Archivio Glottologico Italiano* 83(2). 121–149.
Däbritz, Chris Lasse. 2021. Typology of number systems in languages of Western and Central Siberia. *Finnisch-Ugrische Forschungen* 66. 85–138.
Dali, Myriam & Eric Mathieu. 2021. Singulative systems. In Patricia Cabredo Hofherr & Jenny Doetjes (eds.), *The Oxford handbook of grammatical number*, 275–290. Oxford: Oxford University Press.
Di Garbo, Francesca & Yvonne Agbetsoamedo. 2018. Non-canonical gender in African languages: a typological survey of interactions between gender and number, and between gender and evaluative morphology. In Sebastian Fedden, Jenny Audring & Greville G. Corbett (eds.), *Non-canonical gender systems*, 176–210. Oxford: Oxford University Press.
Dimmendaal, Gerrit J. 2021. A typological perspective on the morphology of Nilo-Saharan languages. In Rochelle Lieber (ed.), *The Oxford encyclopedia of morphology.* Oxford: Oxford University Press.
Dimmendaal, Gerrit J. 2000a. Number marking and noun categorization in Nilo-Saharan languages. *Anthropological Linguistics* 42(2). 214–261.
Dimmendaal, Gerrit J. 2000b. Noun classification in Baale. In Rainer Vossen, Angelika Mietzner & Antje Meissner (eds.), *"Mehr als nur Worte...": afrikanistische Beiträge zum 65. Geburtstag von Franz Rottland*, 183–203. Köln: Rüdiger Köppe Verlag.
Dires, Rahel T. Forth. A morphological analysis on singulatives in Cushitic languages. Accepted to *Linguistique & Langues Africaines.*
Dryer, Matthew S. 2013. Coding of nominal plurality. In Matthew S. Dryer & Martin Haspelmath (eds.), *The world atlas of language structures online.* Leipzig: Max Planck Institute for Evolutionary Anthropology. http://wals.info/chapter/33 (accessed on 27/05/2022)
Fynes-Clinton, Osbert Henry. 1913. *Welsh vocabulary of the Bangor district.* 2 vols. Oxford: Oxford University Press.
Goddard, Cliff. 2010. A piece of cheese, a grain of sand: The semantics of mass nouns and unitizers. In Francis Jeffry Pelletier (ed.), *Kinds, things, and stuff: Mass terms and generics*, 132–165. Oxford: Oxford University Press.
Greenberg, Joseph H. 1963. Some universals of grammar with particular reference to the order of meaningful elements. In Joseph H. Greenberg (ed.), *Universals of language*, 73–113. Cambridge, MA: MIT Press.
Greenberg, Joseph H. 1966. *Language universals: with special reference to feature hierarchies.* The Hague: Mouton. [repr. with foreword by M. Haspelmath, 2005].
Greenberg, Joseph H. 1972. Numeral classifiers and substantial number: Problems in the genesis of a linguistic type. *Stanford papers on language universals* 9. 1–39.
Grimm, Scott. 2012. Individuation and inverse number marking in Dagaare. In Diane Massam (ed.), *Count and mass across languages*, 75–98. Oxford: Oxford University Press.

Grimm, Scott. 2018. Grammatical number and the scale of individuation. *Language* 94(3). 527–574.
Grimm, Scott. 2021. Inverse number in Dagaare. In Patricia Cabredo Hoffher & Jenny Doetjes (eds.), *The Oxford handbook of number in language*, 445–462. Oxford: Oxford University Press.
Gunnink, Hilde. 2018. *A grammar of Fwe: A Bantu language of Zambia and Namibia*. Ph.D. dissertation, University of Gent.
Güldemann, Tom & Jan Junglas. Forth. The four-way meaning of tripartite number: Implications for a typology of number morphology. Unpublished manuscript, Humboldt University Berlin.
Güldemann, Tom. 2018. Historical linguistics and genealogical language classification in Africa. In Tom Güldemann (ed.), *The Languages and Linguistics of Africa*, 58–444. Berlin/Boston: De Gruyter Mouton.
Hammarström, Harald, Robert Forkel, Martin Haspelmath & Sebastian Bank. 2021. *Glottolog 4.4*. Leipzig: Max Planck Institute for Evolutionary Anthropology. https://glottolog.org/ (accessed 23/06/2022).
Haspelmath, Martin. 2010. Comparative concepts and descriptive categories in crosslinguistic studies. *Language* 86(3). 663–687.
Haspelmath, Martin & Andres Karjus. 2017. Explaining asymmetries in number marking: singulatives, pluratives, and usage frequency. *Linguistics* 55(6). 1213–1235.
Haspelmath, Martin (& Edith Moravcsik). 2019. A discussion with Edith Moravcsik about singulative markers and individualizers. *Diversity Linguistics Comment*. Posted on 2019-06-26 by Martin Haspelmath. https://dlc.hypotheses.org/1808 (accessed 09/02/2022).
Helimski, Eugene. 2016. S-singulatives in Ket. *Journal of Language Relationship* 14(3). 157–163.
Jackendoff, Ray. 1991. Parts and boundaries. *Cognition* 41(1–3). 9–45.
Jouitteau, Mélanie. 2009–2022. *ARBRES, wikigrammaire des dialectes du breton et centre de ressources pour son étude linguistique formelle*. IKER, CNRS. http://arbres.iker.cnrs.fr. (accessed 02/02/2022)
Jouitteau, Mélanie & Milan Rezac. 2018. Tester les noms collectifs en breton, enquête sur le nombre et la numérosité. In Ronan Le Coadic (ed.), *Mélanges en l'honneur de Francis Favereau*, 331–364. Morlaix: Skol Vreizh.
Jouitteau, Mélanie & Milan Rezac. 2016. Fourteen tests for Breton collectives, an inquiry into number and numerosity. *Lapurdum* 19. 357–389.
Jurafsky, Daniel. 1996. Universal tendencies in the semantics of the diminutive. *Language* 72(3). 533–578.
Kagan, Olga & Silva Nurmio. Forth. Diminutive or singulative? The suffixes *-in* and *-k* in Russian. In Stela Manova, Laura Grestenberger & Katharina Korecky-Kröll (eds.). *Diminutives across languages, theoretical frameworks and linguistic domains*. Trends in Linguistics. Studies and Monographs [TiLSM] 380. Berlin: De Gruyter Mouton.
Kawachi, Kazuhiro. 2007. *A grammar of Sidaama (Sidamo), a Cushitic language of Ethiopia*. PhD dissertation, University at Buffalo.
Kellerman, Ivy. 2005. *A complete grammar of Esperanto*. Project Gutenberg.
Kossmann, Maarten. 2013a. *A grammatical sketch of Ghadames Berber (Libya)*. (Berber Studies 40). Köln: Rüdiger Köppe.
Kossmann, Maarten. 2013b. *The Arabic influence on Northern Berber*. Leiden/Boston: Brill.
Kossmann, Maarten. 2020. Berber. In Rainer Vossen & Gerrit J. Dimmendaal (eds.), *The Oxford Handbook of African Languages*. Oxford: Oxford University Press.
Ladd, D. Robert, Bert Remijsen & Caguor Adong Manyang. 2009. On the distinction between regular and irregular inflectional morphology: Evidence from Dinka. *Language* 85(3). 659–670.
Laine, Antti. 2023. *Microvariation in Western Serengeti: Comparative morphosyntax of Ikoma, Ishenyi, Nata and Ngoreme*. Ph.D. dissertation. University of Helsinki.

Lehrer, Adrienne. 1986. English classifier constructions. *Lingua* 68(2–3). 109–148.

Le Roux, Pierre. 1924–1953. *Atlas Linguistique de la Basse-Bretagne*. 6 vols. Rennes, Paris: X. http://sbahuaud.free.fr/ALBB/ (accessed 31/01/2022).

Lucy, John A. 1992. *Grammatical categories and cognition: A case study of the linguistic relativity hypothesis*. Cambridge: Cambridge University Press.

Mous, Maarten. 2021. Nominal number in Cushitic. In Patricia Cabredo Hoffher & Jenny Doetjes (eds.), *The Oxford Handbook of Number in Language*, 522–538. Oxford: Oxford University Press.

Nurmio, Silva. 2017. Collective nouns in Welsh: a noun category or a plural allomorph? *Transactions of the Philological Society* 115(1). 58–78.

Nurmio, Silva. 2019. *Grammatical Number in Welsh: Diachrony and Typology*. (Publications of the Philological Society 51). Malden: Wiley Blackwell.

Nurmio, Silva, Rahel T. Dires & Matilda Carbo. Forthcoming. Singulatives: a typological database. In preparation.

Pelletier, Francis Jeffry. 1975. Non-singular reference: some preliminaries. *Philosophia* 5. 451–465.

Plein, Kerstin. 2018. *Verbalkongruenz im Mittelkymrischen*. Hagen: Curach Bhán Publications.

Sandman, Erika. 2016. *A grammar of Wutun*. PhD dissertation. University of Helsinki.

Seifart, Frank. Forth. Towards a typology of unitization: Miraña noun classes compared to numeral classifiers and singulatives. Unpublished Manuscript, Max Planck Institute for Evolutionary Anthropology Leipzig.

Spagnolo, Lorenzo M. 1933. *Bari grammar*. Verona: Missioni africane.

Storch, Anne & Gerrit J. Dimmendaal. 2014. One size fits all? On the grammar and semantics of singularity and plurality. In Anne Storch & Gerrit Dimmendaal (eds.), *Number – Constructions and semantics: Case studies from Africa, Amazonia, India and Oceania*, 1–32. Amsterdam: John Benjamins.

Sutton, Logan. 2010. Noun class and number in Kiowa-Tanoan: Comparative-historical research and respecting speakers' rights in fieldwork. In Andrea L. Berez, Jean Mulder & Daisy Rosenblum (eds.), *Fieldwork and linguistic analysis in indigenous languages of the Americas*, 57–89. (Language Documentation & Conservation Special Publication No. 2). Honolulu: University of Hawai'i Press.

Thomas, R. J., Gareth A. Bevan, Patrick J. Donovan & Andrew Hawke (eds.) 1950–2019. *Geiriadur Prifysgol Cymru/A Dictionary of the Welsh Language*. Cardiff: University of Wales Press, available online at www.geiriadur.ac.uk. (accessed 23/06/2022).

Timyan, Judith E. 1976. *A discourse-based grammar of Baule: Kode dialect*. Ph.D. dissertation, City University of New York.

Treis, Yvonne. 2014. Number in Kambaata: A category between inflection and derivation. In Anne Storch & Gerrit J. Dimmendaal (eds.), *Number – Constructions and semantics: Case studies from Africa, Amazonia, India and Oceania*, 111–133. Amsterdam: John Benjamins.

Trépos, Pierre. 1980. *Grammaire bretonne*. Rennes: Ouest-France.

Trépos, Pierre. 1982. *Le pluriel breton*. 2nd edn. Brest: Emgleo Breiz.

Vajda, Edward. 2022. Number in Ket (Yeniseian). In Paolo Acquaviva & Michael Daniel (eds.), *Number in the world's languages: A comparative handbook*, 307–350. Berlin/Boston: De Gruyter Mouton.

Williams, Stephen J. 1980. *A Welsh grammar*. Cardiff: University of Wales Press.

Wonderly, William L., Lorna F. Gibson & Paul L. Kirk. 1954. Number in Kiowa: nouns, demonstratives, and adjectives. *International Journal of American Linguistics* 20(1). 1–7.

Worku, Firew Girma. 2020. *A grammar of Mursi, a Nilo-Saharan language*. Ph.D. dissertation. James Cook University, Cairns.

Ylikoski, Jussi. 2021. Ice eyes, blood eyes: Remarks on the Uralic singulative marker **ćilmä* 'eye'. Conference poster presentation, *Språkets funktion 17, Turku, 26–27 May 2021*.

Ylikoski, Jussi. 2022. Čalbmi čalmmis ja suoldnečalmmit suoidnečalmmis: Sámegielaid singulatiivvat. *Nordlyd* 46(1). 299–307.

Yue, Anne O. 2016. Chinese dialects: Grammar. In Randy J. LaPolla & Graham Thurgood (eds.), *The Sino-Tibetan Languages*. 2nd edn. London: Routledge.

Zeuss, Johann Caspar. 1853. *Grammatica Celtica*. Leipzig: Weidmann.

Index

https://doi.org/10.1515/9783110986600-006

www.ingramcontent.com/pod-product-compliance
Lightning Source LLC
Chambersburg PA
CBHW051423090426
42737CB00014B/2796